Communications in Computer and Information Science　　1504

More information about this series at https://link.springer.com/bookseries/7899

Yue Xu · Rosalind Wang · Anton Lord ·
Yee Ling Boo · Richi Nayak · Yanchang Zhao ·
Graham Williams (Eds.)

Data Mining

19th Australasian Conference on Data Mining, AusDM
Brisbane, QLD, Australia, December 14–15, 2021
Proceedings

 Springer

Editors
Yue Xu (iD)
Queensland University of Technology
Brisbane, QLD, Australia

Rosalind Wang (iD)
Western Sydney University
Parramatta, NSW, Australia

Anton Lord (iD)
University of Queensland
Herston, Australia

Yee Ling Boo (iD)
RMIT University
Melbourne, VIC, Australia

Richi Nayak (iD)
Queensland University of Technology
Brisbane, QLD, Australia

Yanchang Zhao (iD)
Data61, CSIRO
Canberra, ACT, Australia

Graham Williams (iD)
Australian National University
Canberra, ACT, Australia

ISSN 1865-0929   ISSN 1865-0937 (electronic)
Communications in Computer and Information Science
ISBN 978-981-16-8530-9   ISBN 978-981-16-8531-6 (eBook)
https://doi.org/10.1007/978-981-16-8531-6

This Springer imprint is published by the registered company Springer Nature Singapore Pte Ltd.
The registered company address is: 152 Beach Road, #21-01/04 Gateway East, Singapore 189721, Singapore

Preface

It is our great pleasure to present the proceedings of the 19th Australasian Data Mining Conference (AusDM 2021) held at the Queensland University of Technology, Brisbane, during December 14–15, 2021.

The AusDM conference series first started in 2002 as a workshop initiated by Simeon Simoff (Western Sydney University), Graham Williams (Australian National University), and Markus Hegland (Australian National University). Over the years, AusDM has established itself as the premier Australasian meeting for both practitioners and researchers in area of data mining (or nowadays called data analytics or data science). AusDM is devoted to the art and science of intelligent analysis of (usually big) data sets for meaningful (and previously unknown) insights. Since AusDM 2002, the conference series has showcased research in data mining through presentations and discussions on the state-of-art research and development. Built on this tradition, AusDM 2021 successfully facilitated the cross-disciplinary exchange of ideas, experiences, and potential research directions, and pushed forward the frontiers of data mining in academia, government, and industry.

AusDM 2021 received 32 valid submissions, with authors from 13 different countries. The top three countries, in terms of the number of authors who submitted papers to AusDM 2021, were Australia (81 authors), India (nine authors), and Germany (eight authors). All submissions went through the double-blind review process, and each paper received at least three peer reviewed reports. Additional reviewers were considered for a clearer review outcome, if review comments from the initial three reviewers were inconclusive.

Out of these 32 submissions, a total of 16 papers were finally accepted for publication. The overall acceptance rate for AusDM 2021 was 50%. To further drill into the composition of papers submitted, there were 17 Research Track submissions and eight papers (i.e. 47%) were accepted for publication. On the other hand, out of the 15 submissions in the Application Track, eight papers (i.e. 53%) were accepted for publication.

The AusDM 2021 Organizing Committee would like to give their special thanks to Jeremy Howard and Sharma Chakravarthy for giving the insightful keynote speeches. The committee were appreciative to have Charles Palmer, Alana Maurushat, and Anton Lord presenting and sharing their industry experiences at the Industry Showcase track. The committee would like to give their sincere thanks to Queensland University of Technology for providing administration support and the conference venue. The committee would also like to thank Springer CCIS and the Editorial Board for their acceptance to publish AusDM 2021 papers. This will give the accepted papers excellent exposure and facilitate the dissemination of knowledge through publication. We would also like to give our heartfelt thanks to all student and staff volunteers at the Queensland University of Technology who did a tremendous job in ensuring a successful conference event.

Last but not least, we would like to give our sincere thanks to all delegates for attending the conference this year virtually or in-person at Queensland University of Technology. We hope that it was a fruitful experience and you enjoyed AusDM 2021!

December 2021

Yue Xu
Rosalind Wang
Anton Lord
Yee Ling Boo
Richi Nayak
Yanchang Zhao
Graham Williams

Organization

Conference Chairs

Richi Nayak   Queensland University of Technology, Australia
Yanchang Zhao   Data61, CSIRO, Australia
Graham Williams   Australian National University, Australia

Program Chairs (Research Track)

Yue Xu   Queensland University of Technology, Australia
Rosalind Wang   Western Sydney University, Australia

Program Chair (Application Track)

Anton Lord   Leap in!, Australia

Program Chair (Industry Track)

Warwick Graco   Australian Tax Office, Australia

Organizing Chairs

Khanh Luong   Queensland University of Technology, Australia
Thirunavukarasu   Queensland University of Technology, Australia
  Balasubramaniam

Publicity Chair

Md Abdul Bashar   Queensland University of Technology, Australia

Proceedings Chair

Yee Ling Boo   RMIT University, Australia

Steering Committee Chairs

Simeon Simoff   Western Sydney University, Australia
Graham Williams   Australian National University, Australia

Steering Committee

Peter Christen	Australian National University, Australia
Ling Chen	University of Technology Sydney, Australia
Zahid Islam	Charles Sturt University, Australia
Paul Kennedy	University of Technology Sydney, Australia
Yun Sing Koh	University of Auckland, New Zealand
Jiuyong (John) Li	University of South Australia, Australia
Richi Nayak	Queensland University of Technology, Australia
Kok-Leong Ong	RMIT University, Australia
Dharmendra Sharma	University of Canberra, Australia
Glenn Stone	Western Sydney University, Australia
Yanchang Zhao	Data61, CSIRO, Australia

Honorary Advisors

John Roddick	Flinders University, Australia
Geoff Webb	Monash University, Australia

Program Committee

Research Track

Xuan-Hong Dang	IBM Thomas J. Watson Research Center, USA
Philippe Fournier-Viger	Shenzhen University, China
Yang Gao	Beijing Institute of Technology, China
Md Zahidul Islam	University of South Australia, Australia
Ashad Kabir	Charles Sturt University, Australia
Selasi Kwashie	Data61, CSIRO, Australia
Gang Li	Deakin University, Australia
Wolfgang Mayer	University of South Australia, Australia
Muhammad Marwan Muhammad Fuad	Coventry University, UK
Dang Nguyen	Deakin University, Australia
Quang Vinh Nguyen	Western Sydney University, Australia
Md Anisur Rahman	Charles Sturt University, Australia
Md Geaur Rahman	Charles Sturt University, Australia
Xiaohui Tao	University of Southern Queensland, Australia
Dhananjay Thiruvady	Deakin University, Australia
Bay Vo	Ho Chi Minh City University of Technology, Vietnam
Jie Yang	University of Wollongong, Australia
Weijia Zhang	Southeast University, China

Application Track

Huma Ameer	National University of Sciences and Technology, Pakistan
Ghazal Bargshady	University of Canberra, Australia
Rohan Baxter	Australian Tax Office, Australia
Rushit Dave	North Carolina Agricultural and Technical State University, USA
Stewart Heitmann	Victor Chang Cardiac Research Institute, Australia
Hafsa Ismail	University of Canberra, Australia
Disha Kalal	Leap in!, Australia
Aunsia Khan	National University of Modern Languages, Pakistan
Apurva Khemani	Queensland University of Technology, Australia
Paul Visendi Muhindira	Queensland University of Technology, Australia
Winnie Nakiyingi	African Institute for Mathematical Sciences, Rwanda
Xueping Peng	University of Technology Sydney, Australia
Sharon Torao Pingi	Queensland University of Technology, Australia
Yanfeng Shu	CSIRO, Australia
Hao Zhu	Australian National University, Australia

Additional Reviewers

Rasoul Amirzadch
Fatemah Ansarizadch
Saqib Nawaz
Md Shakil

Contents

Research Track

Parallel Nonlinear Dimensionality Reduction Using GPU Acceleration

Yezihalem Tegegne[1](✉), Zhonglin Qu[1], Yu Qian[2], and Quang Vinh Nguyen[3]

[1] School of Computer, Data and Mathematical Sciences, Western Sydney University, Penrith, Australia
{19201971,18885806}@student.westernsydney.edu.au
[2] J. Craig Venter Institute, La Jolla, CA, USA
mqian@jcvi.org
[3] MARCS Institute and School of Computer, Data and Mathematical Sciences, Western Sydney University, Penrith, Australia
q.nguyen@westernsydney.edu.au

Abstract. Dimensionality reduction is usually an essential step in data mining and classical machine learning from high-dimensional data. Uniform Manifold Approximations Projections (UMAP) is a recently developed nonlinear dimensionality reduction method that is being widely applied in biomedical informatics. However, the UMAP implementation is still not efficient enough for processing the recent big omics data from biomedicine. This paper proposes and implements a method that reduces UMAP runtime using GPU-acceleration on the GPU-RAPIDS platform. Our experiments showed that the parallel UMAP implementation performed hundred times faster than the original UMAP implementation on a cluster computer, while maintaining the effectiveness on identifying leukemic cells from clinical flow cytometry data.

Keywords: Dimensionality reduction · UMAP · GPU-RAPIDS · GPU · cuML · Flow cytometry · Machine learning

1 Introduction

Classical data pre-processing techniques [1] have become slow for big data generated in biomedicine and real world applications, due to both increasing data volume and high data dimensionality. Extracting and visualizing patterns from high-dimensional big data through traditional pairwise comparisons using scatterplots or dot plots can be inefficient and sometimes even infeasible due to curse of dimensionality.

Dimensionality reduction methods project high-dimensional data to low-dimensional data, usually 2D or 3D for straightforward analysis and interpretation of patterns such as data clusters. Linear projection methods, such as Principle Component Analysis (PCA), have been recognized as a basic yet effective method to support high throughput cell biology [2, 3]. While the linear projection approach does not usually capture non-linear relationships which may exist in omics data [4], it can be used

© Springer Nature Singapore Pte Ltd. 2021
Y. Xu et al. (Eds.): AusDM 2021, CCIS 1504, pp. 3–15, 2021.
https://doi.org/10.1007/978-981-16-8531-6_1

as an early process before applying more advanced dimensionality reduction methods [5]. Non-linear dimensionality reduction methods have recently gained popularity in biomedical data analysis due to their ability to identify novel non-linear patterns. T-distributed Stochastic Neighbour Embedding t-SNE [6] (or viSNE [7]) and Uniform Manifold Approximations Projections (UMAP) [8, 9] are among the most commonly used techniques [4, 5]. For example viSNE has been used to identify abnormal T-cell populations in routine clinical flow cytometry data [10]. The recently introduced UMAP has gained popularity due to its fast run time to scale to handle large numbers of cells and dimensions [5, 9].

The non-linear dimensionality reduction methods, such as UMAP are normally implemented to run on Central Processing Units (CPUs) and they are not optimized for parallelization. For example, the UMAP method was originally designed to be compatible with Scikit-learn, NumPy, and Pandas on the CPUs to transfer and visualize data [8]. Scikit-learn [11] is a python module integrating a wide range of state-of-the-art supervised and unsupervised machine learning algorithms that also support data processing and manipulation, such as classification, regression, clustering, model selection. NumPy [12] is an open python module that provides mathematical computation on large multi-dimensional matrices and arrays. And Pandas [13] is an open-source library for data analysis such as reading and writing data between in-memory data structure and different formats such as CSV, text files, and database.

However, these libraries are only support of a portion of their CPU equivalents and work with the data sets that fit in the device memory [14]. For example, running a big data set with Pandas and applying a massive matrix computation with NumPy are time taking. This makes CPU-based dimensionality reduction methods relatively ineffective to process data sets with millions or more items, which limits their capability for real-time or near real-time analysis. Therefore, it is required a new and better mechanism to perform UMAP much faster than the current implementation system to handle such huge omics data sets.

We introduce a new method that optimises the implementation of UMAP algorithm with parallel processing on GPU-RAPIDS platform. Graphical Processing Units (GPUs) are more effective and powerful for parallel processing thanks to their high number of cores per unit in comparison with CPUs. We improve the computational speed significantly which is a hundred times faster or more. While UMAP is a random process and each time the layout is slightly different, our method also preserves similar outputs in comparison with the original CPU-based algorithm.

2   GPU-RAPIDS Platform for UMAP

2.1   GPU-RAPIDS Platform

The advancement of GPU computing is Nvidia's CUDA solution which provides both the CUDA programming language and the enabled hardware computational engine. This platform has a highly parallel architecture with hundreds of cores and very high memory bandwidth, which enables the parallel computation of machine learning models [15]. RAPIDS [16] is an open-source developed by Nvidia and contains APIs and python libraries that allow for executing algorithms python jobs on GPU to speed up machine learning algorithms. And for this, GPU acceleration has become more and more important. In addition, to execute and accelerate tasks using GPU, RAPIDS provides interoperability with other open-source tools for machine learning and deep learning workflows [17]. We use cuDF [18] instead of Pandas, and cuML [19] instead of Scikit-learn to implement UMAP on the GPU-RAPIDS. CuDF is a Python GPU DataFrames for data handling and manipulation. CuML is a Python GPU machine learning library that contains machine learning algorithms that Scikit-Learn has, all in a very similar format but on GPU.

2.2 Constructing k-NN Graph: UMAP-Learn vs CuML-UMAP

Most dimension reduction algorithms can be categorized as either matrix factorization or graph layout. The current implementation system UMAP-Learn uses the nearest neighbours descent algorithm for the construction of an approximate nearest neighbours graph and stochastic gradient descent (SGD) algorithm for optimization. Within our knowledge, UMAP-Learn is only implemented on a single core without the GPU-accelerated version. Tree-based approximate variants, such as algorithms available in Scikit-Learn, also do not have a straightforward port to the GPUs. This is due to the bottleneck of computing extraction of k-NN graph for a large data sample. In order to construct the initial high-dimensional graph, UMAP builds a "fuzzy simplicial complex" graph which represents a weighted graph [8]. The edge weight reflects the likelihood that two nodes are connected to form an edge.

2.3 Get the Nearest Neighbours

A value in a data set is mathematically a vector, and most of the time the rows are made up of numerical values. We calculate the distance between two rows or two vectors to draw a straight line. The straight line distance between two vectors can be calculated by using Euclidean distance formula, which is the square root of the sum of the squared difference between two vectors using the following formula:

$$\sqrt{\sum_i^N (x_{1i} - x_{2i})^2}$$

Where x_1 and x_2 are the first and the second rows of the data respectively, and i is the index to a specific column as we sum across all the columns in the data set and N is the total number of dimensions.

From the distance computation process above, we determine the k-closest instances for the data. We next identify the nearest neighbours in the algorithm. Firstly, we extend a radius outward from each data point to connect other points. When the radii overlap, a weighted k-nearest neighbour (k-NN) graph is constructed. The construction of the graph can be performed via a nearest neighbour or approximate nearest neighbour search algorithm. The computational time for this k-NN graph may take up to 26% of the total time of UMAP-learn [20], which is claimed the second-largest computational bottleneck of the UMAP algorithm, next to the optimization of the embedding process.

2.4 k-NN Graph Optimisation on GPU

The first phase of UMAP algorithm is the construction of a weighted k-NN graph. The bottleneck of a k-NN implementation on a single core is the distance calculation and the selection of k nearest neighbour observation. This problem could be addressed by changing a sequential program into a parallel process, which increases the speed of computation. Implementing k-NN on GPU enables a dramatic computational performance improvement due to the high parallelization nature of the arithmetic operation. Under UMAP-learn k-NN is set to automatically chooses the algorithm based on the size and structure types of the input data set, such as either a brute force search or k-d tree or ball tree search. However, this process is slow and time taking.

The distance calculation process can be fully parallelized on GPU-RAPIDS, hence every thread can involve in the computation of finding the distance between the nearest data points. For large data points, many threads and blocks are launched to perform the computation simultaneously. Once the distance is calculated, the distance kernel is invoked, and the results are stored in shared memory.

The next task is to find the best k-nearest neighbours based on the calculated distance. Then each thread considers one distance simultaneously to calculate and to find the k-nearest neighbours. This first step process saves a significant amount of time when running the UMAP algorithm.

2.5 Embedding Optimisation

Embeddings are robust representations of data modalities like text, images, sound, etc. Essentially, they are vectors of relatively lower dimensions, that can capture original information and it gets a lot easier if we operate on the data using their embeddings representations. The curse of dimensionality complicates research and development works such as in Omics data analytics [21]. UMAP is very efficient at identifying subpopulations in large high dimensional data sets. It scales well with both input dimension and embedding dimension. As mentioned in previous sections, UMAP is comprised of two important steps including first computing a graph representing our data, and second learning an embedding for the graph. The UMAP algorithm uses a binary cross-entropy as a cost function and utilises a stochastic gradient descent to minimise the cost function to optimise the step performance.

The first step of the embedding optimization stage is to initialise the array of output embedding which can follow both random and spectral initialisation strategies. The current implementation uses a spectral embedding of the fuzzy union through the nearest-neighbours variant of the Laplacian eigenmaps algorithm. Our new RAPIDS UMAP method uses spectral clustering implementation from cuGraph, a RAPIDS library that collects GPU accelerated graph algorithms [22]. Overall, we enable parallel processing to perform different UMAP computation steps simultaneously.

UMAP uses binary cross-entropy (CE) as a cost function associated with the model, which can measure how well a model performs on the training data with an aim to minimize the cost function in a way that it comes as close to zero as possible. Gradient descent is an optimization algorithm used to find the values of parameters (coefficients) of a function that minimizes a cost function. Hence gradient descent can be slow to run on very large datasets due to one iteration of the gradient descent algorithm requires a prediction for each instance in the training dataset, it can take a long time when we have many millions of instances. To overcome this problem UMAP uses a modified version of gradient decent called stochastic gradient descent, which calculate the gradient using just a random small part of the observations instead of all of them using batch and this approach can reduce computation time. UMAP computation and update operations have been fused into a single kernel and parallelized so that each thread processes one edge of the fuzzy union. The CUDA kernel is scheduled iteratively for n_epochs which means n times of complete passes of the entire training dataset passing through the training or learning process of the algorithm to compute and apply the gradient updates to the embeddings in each epoch. The dependencies between the vertices in the updating of the gradients make this step non-trivial to parallelize efficiently. This decreases the potential for coalesced memory access and creates the need for atomic operations when applying gradient updates. When the k-NN graph is pre-computed, this step can comprise up to 50% [23] of the remaining time spent in the computation. Therefore, the parallelization of the k-NN graph construction process can speed up the performance of the UMAP algorithm significantly.

3   Benchmark Experimental Comparison

We run our new UMAP algorithm on the RAPIDS-GPU platform using various MINST data sets to evaluate its output quality and its computational time. Our experiments indicated that the new RAPIDS-GPU UMAP method has outperformed the CPU method which is over a hundred times faster while producing comparable outputs (see Table 1).

We present two experiments on the F-MNIST [24] and MNIST [25] digits data sets (collected from https://github.com/zalandoresearch/fashion-mnist) in this paper. Our experiments were carried in the Comet platform, an eXtreme Science and Engineering Discovery Environment (XSEDE) cluster with 24 nodes (https://www.sdsc.edu/support/user_guides/comet.html). Each node uses Intel Xeon E5-2680v3 processors, 128 GB DDR4 DRAM, and 320 GB of SSD local scratch memory. The GPU nodes contain four NVIDIA GPUs each. We only use one node in our computational experiment.

Table 1. Datasets used in experiment and computational time for UMAP methods.

Dataset	Rows	Columns	Running time (in seconds) CPU UMAP	Running time (in seconds) RAPIDS UMAP	Improvement
Fashion Minst	70,000	784	260	1.9	>135 times
Minist (Digits)	70,000	784	240	2.3	>104 times
Omics data sets	1.5 million	16	4000	30	>130 times
	5 million	16	too long	960	

3.1 F-MNIST Data Set

F-MNIST (or Fashion MNIST) [24] is a 10-class dataset of 70,000 28 × 28 pixels grayscale images of fashion items (clothing, footwear, and bags), i.e., 70,000 items with 784 dimensions for classification into 10 categories.

We run the experiments on the existing UMAP method on the CPU and on the new GPU-RAPIDS UMAP method respectively. The UMAP projection aims to identify groups of garments in the two-dimensional space. The vectors of pixels for each image are used as input and the UMAP maps them to the two-dimensional space for each image. Here we can see clearly, the important structural properties of the data have been retained while the known classes have been cleanly pulled apart and isolated. For example, the pants, t-shirts and bags are both retained their shape and internal structure (see blue and red items respectively in Fig. 1).

Figures 1a and 1b show the output from the standard UMAP and from the new GPU-enabled UMAP-RAPIDS methods, respectively. The colours are coded according to the clothing category of the original input (e.g. boots are blue, sneakers are green, sandals are yellow, and trousers are red colours). The visualisation clearly shows that a similar output between the existing UMAP and the new GPU enabled UMAP methods. However, the run time on the GPU method is much faster than the current implementation of UMAP on the CPU. Particularly, the running time of UMAP under GPU-RAPID (Fig. 1b) is 1.9 s in comparison to over 260 s on the original CPU platform.

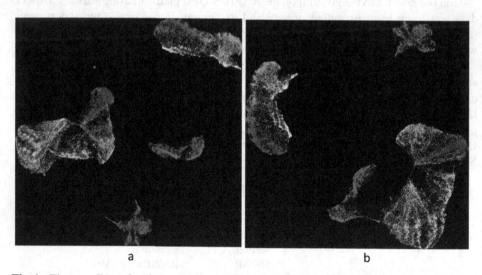

a   b

Fig. 1. The two-dimensional visualisations of UMAP methods on Fashion MNIST dataset, where (a) shows the existing UMAP output (run on CPU) and (b) shows the new UMAP output (run on RAPIDS-GPU platform). It takes over 260 s to generate the output in Fig. 1(a) in comparison with just 1.9 s to generate the output in Fig. 1(b). (Color figure online)

3.2 MNIST Digits Data Set

The MNIST [25] digits data set contains handwritten digits (from 0 to 9) with 70,000 images of digits with 784 dimensions. We use UMAP to identify digits in the data set. After applying UMAP on the data, we find that without any labels, UMAP manages to separate well the data. The outputs are similar between UMAP-learn and UMAP-RAPIDS. Figure 2 shows there are clear 10 clusters that are colour-coded by digit types (0 to 9). However, the performance is much faster for UMAP-RAPIDS (2.3 s) which is more than 100 times faster than the original UMAP (240 s).

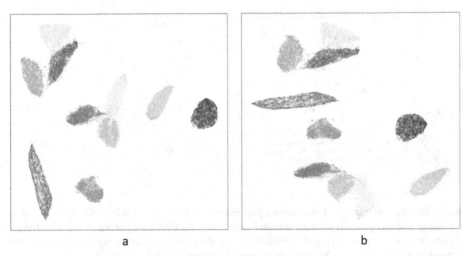

a   b

Fig. 2. The two-dimensional visualisations of UMAP methods on the MNIST digits dataset, where Fig. 2a shows the output of UMAP on CPU and Fig. 2b shows the output of the new UMAP on the RAPIDS-GPU platform. The visualisation shows 10 groups shown in different colours representing 10 digits. It took 240 s to generate the output in Fig. 2a in comparison with just 2.3 s to generate the output in Fig. 2b. (Color figure online)

4 A Case Study on Leukemia Diagnostic Data Analysis

We applied our method to generate UMAP visualization of high-dimensional clinical flow cytometry data for diagnosis of Chronic Lymphocytic Leukemia (CLL). The dataset we used is downloaded from FlowRepository (http://flowrepository.org/) as used and published in a previous study [26]. The goal is to demonstrate the effectiveness of the UMAP dimensionality reduction for supporting visual identification of CLL leukemic cells so that the CLL cases can be visually separated from the non-CLL cases. Each sample contains 10 protein molecule markers and 6 scatter parameters (16 dimensions in total) measured on approximate 100,000 cells (rows). The markers in the reagent panel included CD3, CD5, CD10, CD19, CD22, CD38, CD43, CD45, CD79b and CD81. The scatter attributes include area/height/width for both forward scatter and side scatter parameters (FSC-A/W/H and SSC-A/H/W). To identify CLL cases, we focused on identifying cells with the leukemic phenotype CD5$^+$, CD19$^+$, CD10$^-$, and CD79b^{dim} from the flow cytometry data as defined in the study in [27]. Our analytical workflow of the CLL case studies is illustrated in Fig. 3.

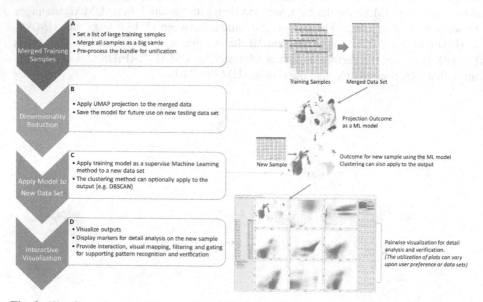

Fig. 3. The flow diagram of the analytics process in the CLL case study includes A) the data processing and merging, B) creating training model(s) with UMAP, C) produce the transformation on the new sample(s) using the saved model, and D) provide interactive visualisation with scatterplots for better comprehension, analysis and verification of the outcome.

We first merged a set of selected CLL samples into one large template for training purpose, resulting in a matrix with the same number of dimensions and the concatenations of millions of rows, with each row representing a cell. We next applied the UMAP non-linear dimensionality reduction method to the merged data. This process projected the 16-dimensional data to a two-dimensional data space for visualization. The template together with UMAP parameters used is then saved for supervised projection of new samples. We repeated this process on different sets and sizes of training samples to generate multiple projection models. This process took most of the computation time since it each time processed a big set of merged data.

For each new diagnostic sample in the analysis, we applied the saved UMAP model as a supervised machine learning process to classify the cellular events onto the 2D UMAP space. A data clustering method, such as DBSCAN [28], is optionally applied to the 2D output for identifying and color-coding of the cell subpopulations. Finally, we implemented interactive visualisation to provide 2D plotting visual presentation on both the UMAP-transformed space and the pairwise combination of original markers with biological meaning side by side for visual pattern recognition and verification. The visualisation utilises the TabuVis tool [29, 30] to provide multiple scatterplots, visual mapping, filtering and interaction enabling effective visual analysis of the data.

We have run the experiments on different sizes of pooled data and training samples. In the first experiment, we carried out the training experiment with 15 CLL samples. After merging all training samples into one large data set (i.e. 1.5 million rows and 16 dimensions), we have applied the UMAP method into the processed data set. The running time on our new UMAP method on the GPU-RAPIDS platform takes approximately 30 s. This is more than a hundred times faster in comparison with over 4000 s when running the same process on a regular CPU system.

In the second experiment, we carried out the training experiment with much larger data sets of 50 CLL samples. We hypothesise that having a much larger number of samples in the training would potentially produce more consistent outcomes and expose better subpopulations in the testing samples. After merging all training samples into a large data set (i.e. 5 million rows and 16 dimensions), we applied the UMAP method into the processed data set. The running time on our new UMAP method on the GPU-RAPIDS platform takes approximately 16 min. Unfortunately, the same process on the CPU system takes multiple days to run so that we had to abort this experiment.

From the saved training models, we can run a new sample independently on a selected model to identify the cancer cells in the data set. This step is much quicker than the training process because we do not need to compute on the much larger merged data. We were able to carry out the computational analysis within a few seconds, which is crucial to enable the real-time diagnostics and analysis of the CLL cancel data. Particularly, our experiments show that the running time GPU-RAPIDS on reading the saved model is 0.4 s for reading a UMAP model with 15 merged samples and less than 1 s for the UMAP model with 50 merged samples respectively. It also takes approximately 3 s for transforming a new CLL data set (100,000 rows and 16 dimensions). The running time for DBSCAN is also less than 1 s. Therefore, the whole computational analysis process takes less than 5 s.

The outcomes of our new UMAP method on GPU-RAPIDS are consistent with the UMAP by CPU method, while the processing time using GPU is more than a hundred times faster than the latter. We can identify quickly the CLL cases as well as the non-CLL cases from the whole dataset in a few minutes. Further from the study in [27], our results are also validated by an domain expert to ensure the accuracy in the diagnosis outcomes of CLL cases.

Figure 4 illustrates the outputs of the projection methods using the training model with 15 CLL samples. We use pairwise scatterplots to show the UMAP output and the selected markers and scatters, where Fig. 4a uses the regular UMAP and Fig. 4b uses our new UMAP method on GPU-RAPIDS. In each figure, the first plot shows the projection output (x-axis and y-axis) and DBSCAN clustering (colours) of the new CLL sample, and other plots show pairwise scatterplots of selected markers and scatter parameters of the new sample, particularly (FSC-A, SSC-A), (SSC-H, CD45), (CD5, CD19), (CD10, CD79b), (CD38, CD22), (CD3, CD81), (CD3, CD19) and (CD79b, CD81) on each x-axis and y-axis respectively. The visualization also highlights the identified CLL cancer cells in red (CD5+, CD19+, CD10−, CD79bdim). The cancer cells are also highlighted simultaneously on other pairwise scatterplots for complete verification and analysis. Both UMAP implementations, as shown in Figs. 4a and 4b, support visual identification of the CLL cells for the new patients effectively.

Figure 5 presents another example of our projection methods using the same training model as in Fig. 4 with 15 CLL samples on a CLL case with high tumour burden, comparing the original UMAP (Fig. 5a) and the GPU-accelerated UMAP implementation on GPU-RAPIDS (Fig. 5b). Both methods can consistently identify the CLL cells for the testing sample. The detailed views show clearly the CLL cells in the 3 pairwise 2D plots (CD5, CD19), (CD10, CD79b) and (CD38, CD22).

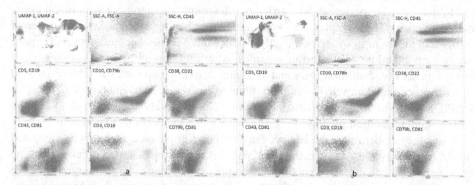

Fig. 4. An example of scatterplots visualizations showing the UMAP projection outcome using the training model with 15 CLL samples. The UMAP outputs are shown in the first panel, and the selected markers and scatters are shown in the other panels. Figures 4a and 4b present the results from the UMAP algorithms on CPU and GPU-RAPIDS respectively.

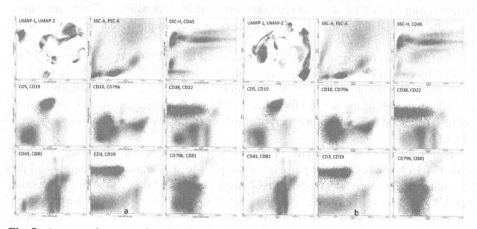

Fig. 5. An example scatterplot visualization showing the UMAP projection on a sample with a large number of cancer cells.

Figure 6 presents the outputs of our UMAP implementation when more training samples were used for the UMAP-based classification. 50 CLL samples were merged to create a UMAP template. The original UMAP implementation was too slow to process such a big input file. Figure 6a indicated 8.9% of CLL cells were found in the testing sample. In Fig. 6b, only a few cells were seen in the UMAP cancer cell region. The results are consistent with the clinical diagnosis labels where the case in Fig. 6a is CLL positive and the case in Fig. 6b non-CLL.

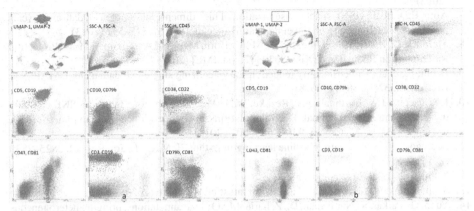

Fig. 6. Two examples of scatterplots visualizations showing the UMAP projection using 50 CLL training samples. Figure 6a show near 8.9% CLL cancer cells in the sample, and Fig. 6b shows a non-CLL case where it shows no cancer cells in the region.

5 Conclusion

We have designed and implemented the UMAP dimensionality reduction on the GPU-RAPIDS platform, which significantly improved the speed of the original UMAP.

We optimised the acceleration and parallel processing with GPUs to overcome to computational bottleneck in the graph construction and distance calculation from the UMAP algorithm. Using MNIST/F-MNIST benchmark datasets, our experiments show that the new method can run a hundred times faster or better, and it also produces stable and consistent outputs in comparison with the original UMAP. We demonstrate the effectiveness of our work with case studies on identifying the cancer cells for clinical diagnosis of leukemia, which will ultimately help real-time application of machine learning diagnosis for biomedicine and the results are also meaningful for other graph and manifold learning algorithms that use GPUs.

References

1. Bendall, S.C., Nolan, G.P., Roederer, M., Chattopadhyay, P.K.: A deep profiler's guide to cytometry. Trends Immunol. **33**, 323–332 (2012)
2. Haghverdi, L., Buettner, F., Theis, F.J.: Diffusion maps for highdimensional single-cell analysis of differentiation data. Bioinformatics **31**, 2989–2998 (2015)
3. Ringnér, M.: What is principal component analysis? Nat. Biotechnol. **26**(3), 303–304 (2008)
4. Konstorum, A., Jekel, N., Vidal, E., Laubenbacher, R.: Comparative analysis of linear and nonlinear dimension reduction techniques on mass cytometry data. bioRxiv 273862 (2018)

5. Luecken, M.D., Theis, F.J.: Current best practices in single-cell RNA-seq analysis: a tutorial. Mol. Syst. Biol. **15**(6), e8746 (2019)
6. Maaten Lvd, Hinton, G.: visualizing data using t-SNE. J. Mach. Learn. Res. **9**, 2579–2605 (2008)
7. Amir, E.D., et al.: viSNE enables visualization of high dimensional single-cell data and reveals phenotypic heterogeneity of leukemia. Nat. Biotechnol. **31**(6), 545–552 (2013)
8. McInnes, L., Healy, J., Melville, J.: UMAP: Uniform Manifold Approximation and Projection for Dimension Reduction. arXiv:180203426 [statML] (2018)
9. Becht, E., et al.: Dimensionality reduction for visualizing single-cell data using UMAP. Nat. Biotechnol. **37**(1), 38–44 (2018)
10. DiGiuseppe, J.A., Cardinali, J.L., Rezuke, W.N., Pe'er, D.: PhenoGraph and viSNE facilitate the identification of abnormal T-cell populations in routine clinical flow cytometric data. Cytometry B Clin. Cytometry **94**(5), 588–601 (2018)
11. Pedregosa, F.: Scikit-learn: machine learning in python. J. Mach. Learn. Res. **12**, 2825–2830 (2011)
12. NumPy (2022). https://numpy.org
13. team Tpd: pandas-dev/pandas: Pandas. In: latest edn: Zenodo (2020)
14. Yuan, G., Palkar, S., Narayanan, D., Zaharia, M.: Offload annotations: bringing heterogeneous computing to existing libraries and workloads. In: Annual Technical Conference (ATC 20), pp. 293–306 (2020)
15. Adu-Gyamfi, Y.: GPU-enabled visual analytics framework for big transportation datasets. J. Big Data Anal. Transp. **1**(2–3), 147–159 (2019). https://doi.org/10.1007/s42421-019-000 10-y
16. RAPIDS: The Platform Inside and Out (2022). https://developer.download.nvidia.com/video/ gputechconf/gtc/(2019)/presentation/s9577-rapids-the-platform-inside-and-out.pdf
17. Aguerzame, A., Pelletier, B., Waeselynck, F.: GPU acceleration of PySpark using RAPIDS AI. In: DATA (2019)
18. Lindholm, E., Nickolls, J., Oberman, S., Montrym, J.: NVIDIA tesla: a unified graphics and computing architecture. IEEE Micro **28**(2), 39–55 (2008)
19. Ocsa, A.: SQL for GPU data frames in RAPIDS Accelerating end-to-end data science workflows using GPUs. In: LatinX in AI Research at ICML (2019)
20. Nolet, C.J., Lafargue, V., Raff, E., Nanditale, T., Oates, T., Zedlewski, J., Patterson, J.: Bringing UMAP Closer to the Speed of Light with GPU Acceleration. arXiv:200800325 [csLG] (2020)
21. Catchpoole, D., Kennedy, P., Skillicorn, D., Simoff, S.: The curse of dimensionality: a blessing to personalized medicine. Proc. Am. Soc. Clin. Oncol. **28**(34), e723–e724 (2010)
22. Hricik, T., Bader, D., Green, O.: Using RAPIDS AI to accelerate graph data science workflows. In: IEEE High Performance Extreme Computing Conference (HPEC), pp. 1–4. IEEE (2020)
23. Nolet, C.J., Lafargue, V., Raff, E., Nanditale, T., Oates, T., Zedlewski, J., Patterson, J.: Bringing UMAP Closer to the Speed of Light with GPU Acceleration (2020)
24. Xiao, H., Rasul, K., Vollgraf, R.J.: Fashion-MNIST: a novel image dataset for benchmarking machine learning algorithms (2017)
25. LeCun, Y., Bottou, L., Bengio, Y., Haffner, P.: Gradient-based learning applied to document recognition. Proc. IEEE **86**(11), 2278–2324 (1998)
26. Ji, D., et al.: machine learning of discriminative gate locations for clinical diagnosis. Cytometry A **97**(3), 296–307 (2020). PMID: 31691488; PMCID: PMC7079150
27. Scheuermann, R.H., Bui, J., Wang, H.-Y., Qian, Y.: Automated analysis of clinical flow cytometry data: a chronic lymphocytic leukemia illustration. Clin. Lab. Med. **37**(4), 931–944 (2017). PMID: 29128077; PMCID: PMC5766345
28. Ester, M., Kriegel, H.-P., Sander, J., Xu, X.: A density-based algorithm for discovering clusters in large spatial databases with noise. In: KDD, pp. 226–231 (1996)

29. Nguyen, Q.V., Qian, Y., Huang, M.L., Zhang, J.: TabuVis: a tool for visual analytics multidimensional datasets. Sci. China Inf. Sci. **56**, 052105:052101–052105:052112 (2013)
30. Nguyen, Q.V., Simoff, S., Qian, Y., Huang, M.L.: Deep exploration of multidimensional data with linkable scatterplots. In: 9th International Symposium on Visual Information Communication and Interaction, pp. 43–50. ACM, Dallas, Texas (2016)

Taking the Confusion Out of Multinomial Confusion Matrices and Imbalanced Classes

David Lovell[1]([✉])[ID], Bridget McCarron[2][ID], Brendan Langfield[3], Khoa Tran[1][ID], and Andrew P. Bradley[1][ID]

[1] Queensland University of Technology, Brisbane, Australia
{David.Lovell,a38.Tran,a6.Bradley}@qut.edu.au
[2] Queensland Health, Brisbane, Australia
Bridget.McCarron@health.qld.gov.au
[3] Services Australia, Brisbane, Australia
Brendan.Langfield@ServicesAustralia.gov.au

Abstract. Classification is a fundamental task in machine learning, and the principled design and evaluation of classifiers is vital to create effective classification systems and to characterise their strengths and limitations in different contexts. Binary classifiers have a range of well-known measures to summarise performance, but characterising the performance of *multinomial classifiers* (systems that classify instances into one of many classes) is an open problem. While *confusion matrices* can summarise the empirical performance of multinomial classifiers, they are challenging to interpret at a glance—challenges compounded when classes are *imbalanced*.

We present a way to decompose multinomial confusion matrices into components that represent the *prior* and *posterior* probabilities of correctly classifying each class, and the intrinsic ability of the classifier to discriminate each class: the *Bayes factor* or *likelihood ratio* of a positive (or negative) outcome. This approach uses the odds formulation of Bayes' rule and leads to compact, informative visualisations of confusion matrices, able to accommodate far more classes than existing methods. We call this method confusR and demonstrate its utility on 2-, 17-, and 379-class confusion matrices. We describe how confusR could be used in the formative assessment of classification systems, investigation of *algorithmic fairness*, and *algorithmic auditing*.

Keywords: Classification · Multiclass · Visualisation · Performance · Fairness · Auditing

1 Introduction

Binary classification systems have a range of performance measures derived from the 2×2 *confusion matrix* produced when a classifier makes predictions about a

---

The original version of this chapter was revised: The formula error in Table 1. has been corrected. The correction to this chapter is available at
https://doi.org/10.1007/978-981-16-8531-6_17

Y. Xu et al. (Eds.): AusDM 2021, CCIS 1504, pp. 16–30, 2021.
https://doi.org/10.1007/978-981-16-8531-6_2

set of examples whose actual classes are known (Table 1). One dimension of the confusion matrix (in this paper, the *rows*) relates to the *predicted* class of each example; the other relates to an example's *actual* class. Some performance measures (e.g., *accuracy, precision, F-score*) depend on the prior abundance of the classes; others—such as the *true* and *false positive rates* from which Receiver Operating Characteristic (ROC) curves are derived—do not [9]. Performance measures that depend on the prior abundance of the classes are especially problematic when classes are *imbalanced* or *skewed*. A trivial example of this is in obtaining an apparent accuracy of 99% from a classifier that always predicts negative, when only 1% of the cases are positive.

Performance measures for binary classifiers are well established in statistics, machine learning and medical decision-making. Not so for *multinomial* classifiers, i.e., systems which classify examples into one of many classes. As organisations seek to develop and deploy these more complex classification systems, there is a growing need for understanding and transparency in model development, as well as a requirement to better understand how the models are operating. This motivates the work that we present here.

Similar to assessment of students' work in educational settings, we can think about performance assessment of classification systems with two ends in mind:

1. *Summative*, in which we wish to have a single measure of performance that we can use to compare and rank different classifiers
2. *Formative*, in which we wish to gain insight into the strengths and limitations of a classification system so we can improve its performance.

We are interested in the latter, noting that in some contexts, understanding and interpretability can trump summative performance: supremely performing models may be blocked from production if they cannot be sufficiently understood. With this in mind, we aim to understand the empirical confusion matrix of a classifier by separating it into components that represent

1. the effect of the prior abundance of different classes in a set of samples presented to the classifier
2. the effect of classifier, i.e., its innate ability to discriminate different classes.

This paper focuses on the visualisation and interpretation of confusion matrices rather than the classification systems that generate them. The design and implementation of multinomial classification systems involves issues such as how to combine the outputs of base classifiers [33], how to set decision thresholds and incorporate misclassification costs [34]. Our hope is that the methods we present here will inform this design and implementation process.

2 Prior Work on Making Sense of Confusion Matrices

We propose a way to visualise the empirical performance of multinomial classification systems using odds ratios to interpret their confusion matrices. Here, we briefly review relevant prior work on these topics.

Table 1. A binary confusion matrix contains the counts of a classifier's predictions in response to a set of examples whose actual classes are known. These counts (TP, FP, FN, TN) are divided by their respective column totals to form the rates TPR, FPR, TNR, FNR. The ratio of true positive rate (TPR) and false positive rate (FPR) is known as the likelihood ratio (or *Bayes factor*) for a positive outcome LR$_+$), and LR$_-$ is defined similarly. The ratio of LR$_+$ and LR$_-$ is known as the *diagnostic odds ratio* (DOR) [11].

predicted class	actual class		
	positive	negative	Pos = TP + FN
			Neg = FP + TN
positive	TP True Positives	FP False Positives	TPR = TP/Pos
			FPR = FP/Neg
			LR$_+$ = TPR/FPR (1)
negative	FN False Negatives	TN True Negatives	TNR = TN/Pos
			FNR = FN/Neg
			LR$_-$ = FNR/TNR
	Pos	Neg	DOR = LR$_+$/LR$_-$ (2)

(a) Binary confusion matrix elements   (b) Binary confusion matrix statistics

"The definition of performance measures in the context of multiclass classification is still an open research topic" remarked Jurman *et al.* [15], citing (then) recent reviews [26], empirical comparisons [10] and visualisation strategies [4] before discussing confusion entropy [30] and a multiclass extension of Matthews correlation coefficient [12] as performance measures. That was in 2012. A more recent investigation suggests the topic remains important and unresolved, and highlights issues with performance indices where classes are imbalanced [20] (see also [18]).

In the machine learning domain, single, *summative* measures for comparison or ranking of multinomial classification systems prevail, with micro- and macro-averaging used to combine performance indices for each class versus all others [26,32]. Cohen's Kappa is also used widely, even though it was not originally intended for classification performance measurement, and has a range of problems when used for that purpose [3]. This extensive use of summative classifier performance metrics may reflect the popularity of competitive evaluation in machine learning, with recognition and reward for those whose algorithms outperform all others—noting that such rankings should be interpreted with care [19]. It also probably reflects the inclusion of these metrics in popular machine learning software frameworks (e.g., [16,21]).

Less common are *formative* approaches that seek to understand, and thereby improve the performance of multinomial classification systems. Ren *et al.* [24] tackle this with their carefully designed and evaluated Squares performance visualization system, providing also a comprehensive review of related visualization efforts such as the Confusion Wheel [1]. Squares is designed to be agnostic to performance metrics and focuses on enabling users to explore calibrated probability scores produced by a classification system in response to test data. Hinterreiter *et al.* [14] propose an interactive system called ConfusionFlow to compare the performances of multinomial classifiers (e.g., during training). In terms of scalability, both Squares and ConfusionFlow were reported to work well with 15–20 classes. Neither approach pays particular attention to class imbalance.

While performance measures like precision, recall, F-score and Area Under the ROC curve (AUC) are popular in machine learning, measures like LR$_+$ (Eq. 1) and DOR (Eq. 1) are not seen so often, even though they are prominent in medical decision-making and diagnosis [5,11,13]. Next, we show how LR$_+$ plays a fundamental role in Bayes' rule that cam be applied to interpreting confusion matrices.

3   Factoring the Confusion Matrix Using Class Odds

Sanderson suggests that we can better understand the discriminative performance of a classification model by expressing Bayes' rule in terms of prior odds and Bayes factors [25]. To illustrate this concept, suppose we have a hypothesis (D) that a person actually has a disease, and some evidence (T) about that in the form of a positive test result for that disease. Often we want to know *"if I have a positive test result, what's the chance that I actually have the disease"*; this is known as the positive predictive value, or precision of the test. In terms of probabilities, this is written:

$$P(D|T) = \frac{P(T|D)P(D)}{P(T|D)P(D) + P(T|\overline{D})P(\overline{D})}$$

where T is the event that the test is positive; D is the event that you actually do have the disease, and $\overline{D}$ is the event that you do not. Sanderson extols the merits of writing this using *odds*:

$$O(D|T) = O(D)\frac{P(T|D)}{P(T|\overline{D})}$$

$$= O(D)\frac{\text{True positive rate}}{\text{False positive rate}}$$

where the ratio is the *Bayes factor* of the test for a positive result, also known as the *likelihood ratio of a positive outcome*, or LR$_+$ for short (Eq. 1). This factor represents how our prior odds of having the disease (O(D)) are updated as a result of the test outcome. In other words

$$\text{posterior odds} = \text{prior odds} \times \text{LR}_+$$

$$= \text{prior odds} \times \frac{\text{True positive rate}}{\text{False positive rate}}$$

In this paper we utilise the realisation that we can visualise these terms on a logarithmic scale, and can exploit the fact that

$$\log(\text{posterior odds}) = \log(\text{prior odds}) + \log(\text{LR}_+) \tag{3}$$

to achieve a graphical presentation in which these odds and LR_+ (the Bayes factor) appear *additively*. This is appealing because these values can be presented in ways that make the most of the human visual system's pre-attentive processing mechanisms [29]. Furthermore, and as we will demonstrate, this strategy can be much more space efficient than the display of 2-dimensional confusion matrices. In reviewing the literature for related work, we learned of Fagan's nomogram [8,13] which is based on similar principles but, as far as we know, has not been used in the interpretation of multinomial confusion matrices.

Since the diagnostic odds ratio (Eq. 2) has a multiplicative relationship with LR_+ and LR_-, we can visualise that additively on a logarithmic scale using the relationship

$$\log(\text{DOR}) = \log(LR_+) + \log(1/LR_-)$$
$$= \log(LR_+) - \log(LR_-). \tag{4}$$

So far, we have used binary classification to illustrate the odds formulation of Bayes' rule. We can extend this to multinomial classification by summarising a $C \times C$ confusion matrix as C binary *one-versus-all* confusion matrices and presenting the prior and posterior odds and Bayes factor of each class against all others. We will demonstrate this approach in the next section, but begin by visualising a well known binary classification scenario that has proven challenging to interpret

4   Application and Demonstration

We have named our confusion matrix visualisation approach confusR because we have implemented it in R [22], benefiting greatly from the tidyverse suite of packages [31]. To highlight the value of our approach We apply confusR to three increasingly challenging confusion matrices.

4.1   Eddy's Probabilistic Reasoning Challenge (2 Classes)

Eddy [6] gives an example of a binary diagnostic test for breast cancer where the test had a TPR of 79.2%, an FPR of 9.6% and the prior probability of breast cancer was assumed to be 1%. Eddy found that most physicians (approximately 95 out of 100 in his informal sample) estimated around a 75% probability of someone actually having cancer given that the test predicted that they had. (What is your estimate?)

Figure 1 shows the confusion matrix we would expect if this test was applied to a sample of 1000 people where the prior probability of cancer was 1% and, for comparison, the confusion matrix expected if the same test was applied to a group in whom the prior probability of cancer was much higher (50%). We then use confusR to visualise Eq. 1, the relationship between the prior and posterior odds and LR_+ (the Bayes factor), for both classes of outcome, *cancer* or *benign*.

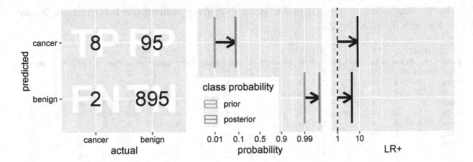

(a) The confusion matrix (and associated statistics) expected when Eddy's diagnostic test [6] is applied to a sample of 1000 people who have a 1% the prior probability of cancer (the red bar at 0.01 in the top row, middle plot). In this scenario, the probability that someone from this group actually has cancer given that the test predicts that they do is just under 10% (the turquoise bar at 0.096 in the top row, middle plot).

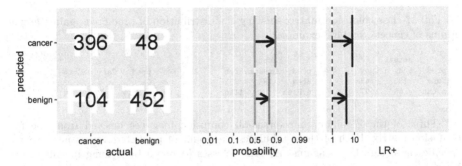

(b) The same test applied to a group of people whose prior probability of cancer is 50% (the red bar at 0.5 in the top row, middle plot). Now the probability that someone from this group actually has cancer given that the test jumps to just under 90% (the turquoise bar at 0.89 in the top row, middle plot).

Fig. 1. Plots of the confusion matrices (left), prior and posterior odds (middle) and likelihood ratios (right) in two different scenarios inspired by the cancer diagnostic test presented by Eddy [6]. Each "row" relates to the *predicted* classes in this binary decision: *cancer* or *benign*. The x-axes of the middle and right plots show the odds and likelihood ratios on a logarithmic scale. (Ticks on the middle plots refer to probabilities for ease of interpretation.) The arrows emphasise that $\log(\text{posterior}) = \log(\text{prior}) + \log(\text{LR}_+)$. Crucially, in both scenarios, the discriminative ability of the test is the same: the right plots of $\log(\text{LR}_+)$ are identical. What differs between the two scenarios is the prior probability of each class (the red bars). On the logarithmic scale used in the middle and right plots, the discriminative ability of the test (LR_+) adds to the prior class odds (red bars) to yield the odds of each class in light of the test's predictions (turquoise bars). (Color figure online)

Table 2. Confusion matrix from Lu *et al.* [17] (Supplementary Information, Source Data Extended Data Fig. 2.)

	actual																
pred	Lung	Brea	Colo	Panc	Skin	Ovar	Rena	Pros	Head	Esop	Thyr	Blad	Germ	Endo	Live	Adre	Cerv
Lung	180	11	0	3	6	2	3	4	3	3	2	3	0	0	2	0	3
Brea	10	194	1	1	3	4	1	0	0	1	2	0	0	1	1	3	3
Colo	3	4	164	1	1	0	0	0	0	2	0	1	0	1	3	1	0
Panc	21	6	6	114	1	2	0	1	2	6	0	2	1	0	0	1	1
Skin	1	4	0	0	90	0	0	1	1	2	0	0	0	0	0	0	0
Ovar	13	5	0	0	2	92	0	0	0	2	0	0	0	5	0	0	0
Rena	0	1	0	0	0	0	70	0	0	0	1	0	2	0	0	0	0
Pros	3	0	0	0	3	0	0	54	0	2	0	0	0	0	0	0	0
Head	1	0	0	0	1	0	2	1	47	0	0	0	0	0	0	0	0
Esop	0	3	2	1	1	0	1	1	3	32	0	1	2	0	0	0	0
Thyr	0	0	0	0	0	0	0	0	0	0	38	0	0	0	0	0	0
Blad	3	2	0	0	3	1	1	2	0	0	0	35	1	0	0	0	0
Germ	0	0	0	0	0	0	0	0	0	1	0	0	26	0	0	0	0
Endo	1	1	0	1	0	1	0	0	0	0	0	0	0	14	7	0	0
Live	0	0	0	1	0	0	1	0	1	0	0	0	0	0	5	0	0
Adre	0	0	0	0	0	0	0	0	0	0	0	0	0	0	0	7	0
Cerv	0	0	2	0	0	0	0	0	0	1	0	0	0	0	0	0	4

(a) Full 17-class confusion matrix showing the distribution of predicted against actual origins of cancers for 1408 examples.

pred	actual			pred	actual			pred	actual	
	Lung	n.Lung			Brea	n.Brea	...		Cerv	n.Cerv
Lung	180	45		Brea	194	31	...	Cerv	4	3
n.Lung	56	1127		n.Brea	37	1146	...	n.Cerv	7	1394

(b) Three of the 17 binary one-versus-all confusion matrices derived from the full confusion matrix. Each of these tabulates the count of predicted versus actual for a specific class (e.g., Lung) against all other classes (denoted by n.Lung meaning "not Lung").

The middle and right panels of Fig. 1 clearly delineate the contribution of the prior probability of each class, the contribution of the classifier's ability to discriminate each class, and the posterior probability of an case actually being from a given class, given that the test's prediction.

Now let us consider a situation with more than two classes.

4.2   Cancer of Unknown Primary—CUP (17 Classes)

Lu *et al.* recently developed and evaluated a deep-learning-based system to classify the origin of a cancer primary tumour from histopathology images [17]. Table 2a shows a 17-class confusion matrix from this study and Table 2b shows three of the 17 binary one-versus-all confusion matrices we can derive from it. This number of classes is towards the upper range of what Squares [24] and ConfusionFlow [14] are designed to represent.

Figure 2a shows a confusR visualisation of the CUP confusion matrix, sorted, by the LR_+, prior and posterior values of the classes. Classes Thyr and Adre stand out with infinite LR_+ and posterior probabilities, signified by bars at the

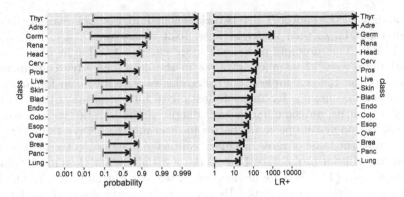

(a) Classes sorted by LR$_+$ (black, right), with prior (red) and posterior (turquoise) probabilities on the left.

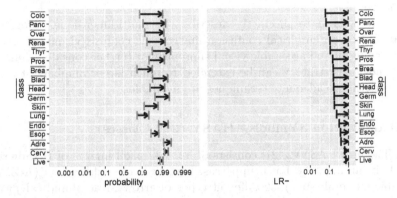

(b) Likelihood ratios of a *negative* result for each class, with classes sorted by LR$_-$ (black, right) and prior (red) and posterior (turquoise) probabilities on the left.

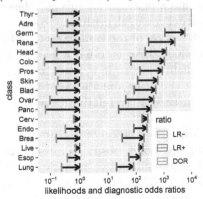

(c) Classes sorted by diagnostic odds ratios (blue) using the relationship in Equation 4. The arrows emphasise that $-\log(\mathrm{LR}_-)$ is added to $\log(\mathrm{LR}_+)$ to get $\log(\mathrm{DOR})$.

Fig. 2. confusR plots of CUP class prior and posterior probabilities, Bayes factors (LR$_+$, LR$_-$) and diagnostic odds ratios. (Color figure online)

extreme right of the panels. These two classes had no false positives in the CUP confusion matrix, so their apparent false positive rate and hence denominator of Eq. 1, is 0. This prompts us to look more closely at Table 2a where we can see that there were 12 instances of `Adre` in the data (the `Adre` column total), and none of the 1396 examples from other classes were mistaken for that class. This is a similar outcome for `Thyr` (43 instances) but, in terms of the likelihood ratio of a negative outcome and overall diagnostic odds ratio, we can see that the classification system performs better in discriminating `Thyr` than `Adre`. We will return to the issue of zeros in binary confusion matrices in the next section.

In terms of prior probability, `Lung` is the most abundant class in this dataset (236 instances). (This is more obvious when we sort Fig. 2a by LR_-, which we omit to stay within page limits.) However, it is poorly discriminated by the classification system, having the lowest LR_+ and DOR. This suggests to us that this class merits more attention than other classes in efforts to improve performance.

While it is hard to succinctly describe the multidimensional information encapsulated by an empirical confusion matrix, the confusR visualisations provide a meaningful and accessible visual summary for further consideration. Next we show how this strategy can be extended to much larger numbers of classes, where currently no adequate techniques are available.

4.3   HAndwritten SYmbols—HASY (379 Classes)

Martin Thoma's HASYv2 [27] consists of 32×32 pixel images of HAndwritten SYmbols including "the Latin uppercase and lowercase characters (A-Z, a-z), the Arabic numerals (0–9), 32 different types of arrows, fractal and calligraphic Latin characters, brackets and more", collected from https://detexify.kirelabs. org/classify.html and http://write-math.com/. Thoma has also developed classifiers and published confusion matrices for this 369 class problem available from https://github.com/MartinThoma/algorithms.

Table 3 shows a small section of the training set confusion matrix in numeric form. One way to see all 369 classes is to use a grey scale image or heat map as in Fig. 3a. But with so many classes, it is hard to get a sense of how well the classifier is doing or what the prior abundance of the classes are. Furthermore, humans have difficulty in accurately relating grey scale or colour intensity to quantity [29, p. 168].

In contrast, the confusR visualisation of Fig. 3b makes the most of our visual system's ability to compare point positions pre-attentively, especially when ordering is used to reduce uninformative variation. In this Figure, classes are ordered from bottom to top by LR_+ then prior probability, allowing us to discern some interesting relationships on this training data. As LR_+ increases (from classes \sum up to \n) we see a rough but noticeable decrease in the prior abundance of classes (red dots). This indicates to us that many of these relatively

Table 3. Section of HASYv2 training set confusion matrix.

predict	actual									
	\\nu	\\xi	\\Xi	\\Pi	\\rho	\\varrho	\\tau	\\phi	\\Phi	\\varphi
\\nu	344	0	0	0	0	0	0	0	0	0
\\xi	0	2309	0	0	0	0	0	0	0	0
\\Xi	0	0	350	0	0	0	0	0	0	0
\\Pi	0	0	0	451	0	0	0	0	0	0
\\rho	0	0	0	0	622	1	0	0	0	0
\\varrho	0	0	0	0	0	198	0	0	0	0
\\tau	0	0	0	0	0	0	369	0	0	0
\\phi	0	0	0	0	0	0	0	561	13	2
\\Phi	0	0	0	0	0	0	0	25	532	0
\\varphi	0	0	0	0	0	0	0	2	0	1366

infrequent classes are distinctive to the classifier. There is also a band of classes from \multmap to \n in which we can clearly see the impact of prior abundance on posterior probabilities while the classifier's ability to discriminate these classes (LR_+) stays steady.

This is a good point to stress that we see these confusR visualisations as *complementary to* the traditional confusion matrix representation rather than a replacement for it. The confusR representation does not show what classes are being confused, but it does give rapid insight into the extent to which classes are being confused, as well as meaningfully factoring apart the role of prior abundance from the classifier's intrinsic ability to discriminate a particular class. We think that this has potential to help developers focus in on more manageable *subsets* of the confusion matrix or data for further attention, perhaps using interactive strategies like Squares [24] or ConfusionFlow [14].

We also see potential for confusR visualisations to compare the empirical performance of a classifier on different datasets, as shown in Fig. 4a which compares confusion matrices from HASYv2 training and test data. The test set confusion matrix has a number of classes for that show zero true positives and/or zero false positives (Fig. 4b), in which case LR_+ is off the scale of Fig. 4a or undefined (when both TP and FP are zero). Figure 4a visualises confusion matrices on two different data sets; this approach could easily be extended to show the distribution of LR_+ values observed across many data sets, e.g., as would occur in cross-validation of classifier performance.

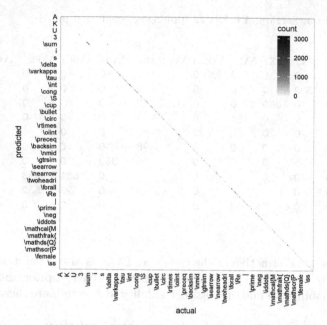

(a) HASYv2 training set confusion matrix as an image in which the count of each element is shown in grey scale.

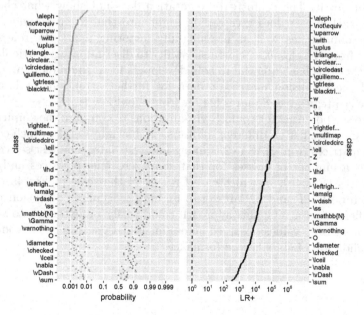

(b) HASY as a confusR one-vs-all plot

Fig. 3. Two ways of visualising HASYv2 training set confusion matrix from https:// github.com/MartinThoma/algorithms. For legibility, only every tenth class label is printed on the y-axes. Classes sorted by LR_+ (black, right), then prior (red) probability with posterior probability in turquoise. (Color figure online)

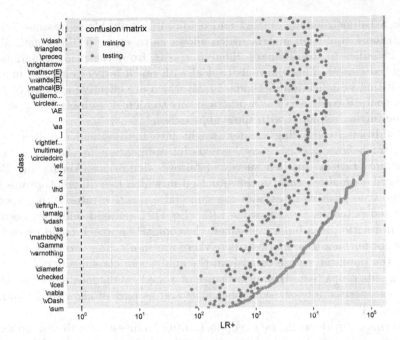

(a) The performance of Martin Thoma's classifier on the HASYv2 training set and test set (from https://github.com/MartinThoma/algorithms). Classes are sorted in order of the empirical LR$_+$ observed on the training set, highlighting the familiar tendency of classifiers to fit the training data best.

	TP $==$ 0	TP $>$ 0
FP $==$ 0	3 LR$_+$ = NaN	34 LR$_+$ = ∞
FP $>$ 0	12 LR$_+$ = 0	320

(b) Confusion matrices may contain classes for which no true positives (TP) or false positives (FP) are observed, giving rise to LR$_+$ values that are zero, infinite, or not defined (NaN: Not a Number). This table shows the number of classes where these LR$_+$ values occurred in the confusion matrix derived from the HASYv2 test set.

Fig. 4. confusR visualisations can be used to compare a classifier's performance on different datasets.

5   Discussion and Conclusions

The odds formulation of Bayes' rule has been around for long time [7], but it does not seem to have been used in visualising the relationship between prior and posterior odds as we have described here with our confusR approach. By putting prior and posterior class odds and Bayes factors onto a logarithmic scale, we provide a 1-dimensional, compact and readily interpretable representation of a 2-dimensional confusion matrix which allows us to separate the innate ability of the classifier to discriminate different classes from the prior abundance of imbalanced classes. This allows us to deal with much larger confusion matrices than existing visualisation methods [14,24] which are effective with up to 15–20 classes; confusR could serve as a practical way to identify subsets of classes that could be further explored with these approaches.

In addition to enabling greater insight into multinomial classifier performance, we believe confusR may have useful application in the evaluation of *algorithmic fairness*. In their survey of fairness definitions, Verma and Rubin [28] describe *equalized odds* as the situation in which two different groups have the same true positive rates, and the same false positive rates with respect to a predicted outcome, i.e., the same LR_+ (Eq. 1). Equalised odds has been discussed in relation to a binary classifier; the confusR approach could extend this measure of fairness to scenarios where there are more than two classes at play. We see the potential to use confusR in the analysis of classifier performance on different subgroups (e.g., the performance of a medical diagnostic for classifying skin lesions across different skin tones [35]) and as part of algorithmic auditing processes [23].

In terms of future work, we see opportunities to investigate the incorporation of uncertainty (e.g., through simulation or theoretical approaches [2]) into the confusR visualisation approach. Incorporation of decision costs [34] would be a valuable advance, but we suspect this would require more information (i.e., calibrated class probability estimates) than empirical confusion matrices alone could provide.

We are developing an R package that implements the methods presented here, however, the underlying computations are simple and we hope that this paper will provide sufficient information for others to use the confusR concept in making sense of confusion matrices and imbalanced classes.

References

1. Alsallakh, B., Hanbury, A., Hauser, H., Miksch, S., Rauber, A.: Visual methods for analyzing probabilistic classification data. IEEE Trans. Visual Comput. Graphics **20**(12), 1703–1712 (2014). https://doi.org/10.1109/TVCG.2014.2346660
2. Caelen, O.: A Bayesian interpretation of the confusion matrix. Ann. Math. Artif. Intell. **81**(3), 429–450 (2017). https://doi.org/10.1007/s10472-017-9564-8
3. Delgado, R., Tibau, X.A.: Why Cohen's Kappa should be avoided as performance measure in classification. PLoS ONE **14**(9), e0222916 (2019). https://doi.org/10.1371/journal.pone.0222916
4. Diri, B., Albayrak, S.: Visualization and analysis of classifiers performance in multiclass medical data. Expert Syst. Appl. **34**(1), 628–634 (2008). https://doi.org/10.1016/j.eswa.2006.10.016
5. Dujardin, B., Van den Ende, J., Van Gompel, A., Unger, J.P., Van der Stuyft, P.: Likelihood ratios: a real improvement for clinical decision making? Eur. J. Epidemiol. **10**(1), 29–36 (1994). https://doi.org/10.1007/BF01717448

6. Eddy, D.M.: Probabilistic reasoning in clinical medicine: Problems and opportunities. In: Tversky, A., Kahneman, D., Slovic, P. (eds.) Judgment under Uncertainty: Heuristics and Biases, pp. 249–267. Cambridge University Press, Cambridge (1982). https://doi.org/10.1017/CBO9780511809477.019

7. Etz, A., Wagenmakers, E.J.: J. B. S. Haldane's contribution to the bayes factor hypothesis test. Stat. Sci. **32**(2), 313–329 (2017)

8. Fagan, T.: Nomogram for bayes's theorem. N. Engl. J. Med. **293**(5), 257–257 (1975). https://doi.org/10.1056/NEJM197507312930513

9. Fawcett, T.: An introduction to ROC analysis. Pattern Recogn. Lett. **27**(8), 861–874 (2006). https://doi.org/10.1016/j.patrec.2005.10.010

10. Ferri, C., Hernández-Orallo, J., Modroiu, R.: An experimental comparison of performance measures for classification. Pattern Recogn. Lett. **30**(1), 27–38 (2009). https://doi.org/10.1016/j.patrec.2008.08.010

11. Glas, A.S., Lijmer, J.G., Prins, M.H., Bonsel, G.J., Bossuyt, P.M.M.: The diagnostic odds ratio: a single indicator of test performance. J. Clin. Epidemiol. **56**(11), 1129–1135 (2003). https://doi.org/10.1016/S0895-4356(03)00177-X

12. Gorodkin, J.: Comparing two K-category assignments by a K-category correlation coefficient. Comput. Biol. Chem. **28**(5), 367–374 (2004). https://doi.org/10.1016/j.compbiolchem.2004.09.006

13. Grimes, D.A., Schulz, K.F.: Refining clinical diagnosis with likelihood ratios. The Lancet **365**(9469), 1500–1505 (2005). https://doi.org/10.1016/S0140-6736(05)66422-7

14. Hinterreiter, A., et al.: ConfusionFlow: a model-agnostic visualization for temporal analysis of classifier confusion. IEEE Trans. Visualization Comput. Graph., 1 (2020). https://doi.org/10.1109/TVCG.2020.3012063

15. Jurman, G., Riccadonna, S., Furlanello, C.: A comparison of MCC and CEN error measures in multi-class prediction. PLoS ONE **7**(8) (2012). https://doi.org/10.1371/journal.pone.0041882

16. Kuhn, M.: Building predictive models in r using the caret package. J. Stat. Softw. Articles **28**(5), 1–26 (2008). https://doi.org/10.18637/jss.v028.i05

17. Lu, M.Y., et al.: AI-based pathology predicts origins for cancers of unknown primary. Nature **594**(7861), 106–110 (2021). https://doi.org/10.1038/s41586-021-03512-4

18. Luque, A., Carrasco, A., Martín, A., de las Heras, A.: The impact of class imbalance in classification performance metrics based on the binary confusion matrix. Pattern Recogn. **91**, 216–231 (2019). https://doi.org/10.1016/j.patcog.2019.02.023

19. Maier-Hein, L., Eisenmann, M., Reinke, A., Onogur, S., Stankovic, M., Scholz, P., et al.: Why rankings of biomedical image analysis competitions should be interpreted with care. Nat. Commun. **9**(1), 5217 (2018). https://doi.org/10.1038/s41467-018-07619-7

20. Mullick, S.S., Datta, S., Dhekane, S.G., Das, S.: Appropriateness of performance indices for imbalanced data classification: an analysis. Pattern Recogn. **102** (2020). https://doi.org/10.1016/j.patcog.2020.107197

21. Pedregosa, F., Varoquaux, G., Gramfort, A., Michel, V., Thirion, B., Grisel, O., et al.: Scikit-learn: machine learning in python. J. Mach. Learn. Res. **12**, 2825–2830 (2011)

22. R Core Team: R: A language and environment for statistical computing. Technical report, Vienna, Austria (2020). https://www.R-project.org/, R Foundation for Statistical Computing

23. Raji, I.D., Smart, A., White, R.N., Mitchell, M., Gebru, T., Hutchinson, B., et al.: Closing the AI accountability gap: defining an end-to-end framework for internal algorithmic auditing. In: Proceedings of the 2020 Conference on Fairness, Accountability, and Transparency, pp. 33–44. ACM, Barcelona (2020). https://doi.org/10.1145/3351095.3372873
24. Ren, D., Amershi, S., Lee, B., Suh, J., Williams, J.D.: Squares: supporting interactive performance analysis for multiclass classifiers. IEEE Trans. Visual Comput. Graphics **23**(1), 61–70 (2017). https://doi.org/10.1109/TVCG.2016.2598828
25. Sanderson, G.: The medical test paradox: can redesigning Bayes rule help? (2020). https://www.youtube.com/watch?v=lG4VkPoG3ko
26. Sokolova, M., Lapalme, G.: A systematic analysis of performance measures for classification tasks. Inf. Process. Manage. **45**(4), 427–437 (2009). https://doi.org/10.1016/j.ipm.2009.03.002
27. Thoma, M.: The HASYv2 dataset. arXiv:1701.08380 [cs] (2017)
28. Verma, S., Rubin, J.: Fairness definitions explained. In: Proceedings of the International Workshop on Software Fairness, FairWare 2018, pp. 1–7. ACM, New York (2018). https://doi.org/10.1145/3194770.3194776
29. Ware, C.: Information visualization: perception for design. Interactive technologies, 3rd edn. Morgan Kaufmann, Waltham (2013)
30. Wei, J.M., Yuan, X.J., Hu, Q.H., Wang, S.Q.: A novel measure for evaluating classifiers. Expert Syst. Appl. **37**(5), 3799–3809 (2010). https://doi.org/10.1016/j.eswa.2009.11.040
31. Wickham, H., et al.: Welcome to the tidyverse. J. Open Source Softw. **4**(43), 1686 (2019). https://doi.org/10.21105/joss.01686
32. Wu, X.Z., Zhou, Z.H.: A unified view of multi-label performance measures. arXiv:1609.00288 [cs] (2017)
33. Zadrozny, B., Elkan, C.: Transforming classifier scores into accurate multiclass probability estimates. In: Proceedings of the Eighth ACM SIGKDD International Conference on Knowledge Discovery and Data Mining, KDD 2002, pp. 694–699. Association for Computing Machinery, New York (2002). https://doi.org/10.1145/775047.775151
34. Zhou, Z.H., Liu, X.Y.: On multi-class cost-sensitive learning. In: Proceedings of the 21st National Conference on Artificial Intelligence, AAAI 2006, vol. 1, pp. 567–572. AAAI Press, Boston (2006)
35. Zicari, R.V., Ahmed, S., Amann, J., Braun, S.A., Brodersen, J., et al.: Co-design of a trustworthy AI system in healthcare: deep learning based skin lesion classifier. Front. Hum. Dyn. **3**, 40 (2021). https://doi.org/10.3389/fhumd.2021.688152

Sharpshooting Most Beneficial Part of AUC for Detecting Malicious Logs

Taishi Nishiyama$^{(\boxtimes)}$, Atsutoshi Kumagai, Akinori Fujino,
and Kazunori Kamiya

NTT Laboratories, 3-9-11 Midori-cho, Musashino-shi, Tokyo 180-8585, Japan
{taishi.nishiyama.pt,atsutoshi.kumagai.ht,akinori.fujino.yh,
kazunori.kamiya.ew}@hco.ntt.co.jp

Abstract. Machine learning is becoming a vital component in cyber security to automatically analyze network logs and detect malware infection. Since actual network logs are imbalanced data that contain fewer malicious logs than benign logs, many studies have evaluated the classification performance of malicious log detection by using the area under the curve (AUC). However, in actual operation, a low false positive rate (FPR) is required to reduce the burden on network operators. Therefore, the area to be focused on is not the entire AUC but the partial AUC (pAUC) in a specific low FPR interval. We propose a novel supervised learning method that maximizes the pAUC, called Partial AUC Utilization for Malicious log Analysis (PUMA). Once the FPR that operators can permit is specified, PUMA optimizes the detection model to obtain the best TPR for the specified FPR interval. In addition, we describe the theoretical formulation and algorithm of PUMA to handle ties in the scores of training data that frequently occur in log analysis and often make existing pAUC maximization methods unsuitable. We also demonstrate the effectiveness of PUMA after comparing it with conventional methods with proxy logs from a real-world large enterprise network.

Keywords: Partial AUC · Malware detection · Log analysis

1 Introduction

Malware has been the primary cyber threat for years. To prevent malware infections, antivirus products, e.g., antivirus software, traditionally use a signature-based detection method [1]. However, cybercriminals continuously devise sophisticated techniques to attack victims more efficiently and escape investigation by security vendors and authorities. Therefore, signatures quickly become obsolete, and infections are overlooked. As a countermeasure, network logs are analyzed to quickly detect malicious activities. To save on labor and develop a more sophisticated detection method, log analysis with machine learning is becoming vital. Even if a piece of malware is new, most of it is not completely new since cybercriminals often repackage parts of the source code of known malware [2].

© Springer Nature Singapore Pte Ltd. 2021
Y. Xu et al. (Eds.): AusDM 2021, CCIS 1504, pp. 31–46, 2021.
https://doi.org/10.1007/978-981-16-8531-6_3

A machine-learning-based detection method finds such similarities between new and known malware and detects suspicious logs from new types of malware.

Many studies have proposed machine-learning-based classifiers for detecting malicious content or malware infection [3–9]. To evaluate the performance of machine-learning-based classifiers, various types of performance measures have been used, e.g., accuracy, true positive rate (TPR), false positive rate (FPR), recall, and precision. However, these performance measures are not appropriate when a dataset contains an imbalance between the number of malicious and benign logs. For example, suppose there are 99 benign logs and 1 malicious log. If the classifier determines that all data are benign, accuracy becomes 99%. Though accuracy is high, it is not practical since it does not detect any malicious logs. Therefore, accuracy is not an appropriate performance measure when the data are imbalanced. On the other hand, the area under the receiver operating characteristic (ROC) curve (AUC) can be used to evaluate imbalanced data since it depends on two types of error rates: FPR and false negative rate (FNR). The ROC curve is obtained by plotting the TPR $(= 1 - \text{FNR})$ against the FPR when varying a classifier's threshold. Since actual network logs are imbalanced data that contain fewer malicious logs than benign logs, the AUC is a good performance measure to evaluate a classifier for malicious log detection.

However, in actual operation, an extremely low FPR is needed, e.g., less than 0.1%, since false positives will impose a heavy burden on network operators. In many cases, network operators have to manually analyze and judge whether a suspicious log detected with a classifier is truly a trace of a cyberattack. Therefore, the most important performance measure in actual operation is not the AUC but the TPR in a low FPR interval. In fact, some researchers use the TPR in the low FPR as a performance measure to evaluate the effectiveness of the classifier [3–6]. For the reasons above, we focus on the partial AUC (pAUC) in a low FPR interval. The pAUC corresponds to the mean TPR in the desired FPR interval. Note that some researchers prefer to use the precision-recall curve rather than the ROC curve when evaluating imbalanced data. Training a classifier to increase the TPR in a low FPR interval, i.e., pAUC, is equivalent to training a classifier to increase recall as much as possible while maintaining a large precision. Therefore, the essence is the same.

In this paper, we propose a novel supervised learning method that maximizes the pAUC, called Partial AUC Utilization for Malicious log Analysis (PUMA). PUMA uses the pAUC formulation as the objective function to be maximized and optimizes the detection model to obtain the best TPR for a suitable range of FPR interval, i.e., pAUC, whereas the conventional supervised learning methods, e.g., logistic regression (LR) and support vector machine (SVM), are trained to maximize accuracy. Once the FPR that network operators can permit is specified in malicious log detection, PUMA increases the score of malicious logs and lowers that of benign logs in the specified FPR interval to maximize the pAUC, assuming that we give a high score to suspicious logs. In addition, we describe the theoretical formulation and algorithm of PUMA to handle ties in the scores of training data scored by the classifier that frequently occur in log analysis due to the effect of the load balancer, etc. Details are described in Sect. 3. Although the number of previous studies on pAUC maximization methods is limited, several researchers in machine learning have proposed methods for directly maximizing the pAUC [10–13]. However, these studies assumed that there are no ties. Therefore, if a tie occurs, previous pAUC maximization methods cannot correctly calculate the pAUC and maximize the pAUC well. Thus, if applied at the same FPR, PUMA can detect more malicious logs than previous pAUC maximization and conventional supervised learning methods.

In addition, to evaluate the effectiveness of PUMA, we compared PUMA with previous pAUC maximization methods and conventional supervised learning methods by using proxy logs from an actual large enterprise network.

In summary, we make the following contributions:

- To the best of our knowledge, this paper is the first to focus on maximizing the pAUC in log analysis and cybersecurity.
- We propose a novel supervised learning method called PUMA to directly maximize the pAUC considering ties in the scores of training data.
- We demonstrated the effectiveness of PUMA through comparative experiments with proxy logs from a real-world large enterprise network.

2   Definition of AUC and pAUC

This section describes how to define the AUC and pAUC. We consider a binary classification task for classifying each log as malicious ("+") or benign ("−").

We prepared two labeled datasets: a malicious dataset $S^+ = \{x_1^+, x_2^+, \cdots, x_m^+\}$ and benign dataset $S^- = \{x_1^-, x_2^-, \cdots, x_n^-\}$, where $x_p \in \mathbb{R}^D$ is a D-dimensional feature vector of a p-th data point, and $\mathbb{R}$ is the set of real numbers. Generally, there are far fewer malicious logs than benign logs, $m \ll n$. We use $f(x; w)$ to denote a scoring function whose parameter vector is w and $t \in \mathbb{R}$ to denote a threshold. That is, a data point p is classified as malicious when $f(x_p; w) > t$ and benign when $f(x_p; w) \leq t$. Assuming there are no ties, the AUC and pAUC of f between a specific FPR interval $[\alpha, \beta]$ ($0 \leq \alpha < \beta \leq 1$) are given by [10–12]

$$\text{AUC}_f = \frac{1}{mn} \sum_{i=1}^{m} \sum_{j=1}^{n} I(f(x_i^+; w) > f(x_j^-; w)), \tag{1}$$

$$\text{pAUC}_f(\alpha, \beta) = \frac{1}{mn(\beta - \alpha)} \sum_{i=1}^{m} \Big[(j_\alpha - n\alpha) \cdot I(f(x_i^+; w) > f(x_{(j_\alpha)}^-; w)) \tag{2}$$

$$+ \sum_{j=j_\alpha+1}^{j_\beta} I(f(x_i^+; w) > f(x_{(j)}^-; w)) + (n\beta - j_\beta) \cdot I(f(x_i^+; w) > f(x_{(j_\beta+1)}^-; w)) \Big],$$

where $j_\alpha = \lceil n\alpha \rceil$, $j_\beta = \lfloor n\beta \rfloor$, $I(z)$ ($\forall z \in \mathbb{R}$) is a Heaviside step function; $I(z) = 1$ if z is true, and $I(z) = 0$ otherwise, and $x_{(j)}^-$ denotes the benign log ranked in the j-th position among S^- in descending order of scores by f. The notation $\lceil * \rceil$ is the smallest integer not less than $*$, and $\lfloor * \rfloor$ is the largest integer no more than $*$. Equation (2) takes the values from 0 to 1 since it is normalized. Note that if $\alpha = 0$ and $\beta = 1$, $\mathrm{pAUC}_f(\alpha, \beta)$ is equivalent to AUC_f.

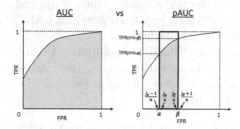

Fig. 1. Difference between the AUC and pAUC. Symbols are given in (2).

Fig. 2. Example of the ROC curve with ties. There are ties where the score is 0.7.

Figure 1 shows the difference between the AUC and pAUC. $\mathrm{TPR}_{\mathrm{FPR}=X}$ is defined as the TPR when adjusting the threshold so that the FPR $= X$. Note that the scoring function with a high AUC is not necessary when we focus on the pAUC. Therefore, many previous studies in security have compared the classification performance with the AUC, which is not always correct in terms of actual operation, since the TPR in the low FPR is an important measure to reduce operational cost. This paper focuses on not the AUC but the pAUC in the low FPR range and proposes a new pAUC maximization method.

3 Problem

Using (2) as a part of the objective function, the conventional pAUC maximization methods [10–13] assume that there are no ties in the scores of training data scored by the classifier. If there is a tie, (1) and (2) have large discrepancies from the actual AUC and pAUC, respectively. Therefore, these equations are inappropriate to use as part of the objective function.

Figure 2 shows an example of the ROC curve with ties. The part of the ROC curve where the score is a tie is a diagonal line. $11/20 = 0.55$ should be estimated as the actual AUC since the estimated AUC fluctuates between $8/20 = 0.4$ and $14/20 = 0.7$ by the order of five examples where the score is 0.7 and the average AUC is 0.55. On the other hand, when we calculate the AUC in accordance with the definition of (1) or (2), AUC is $8/20 = 0.4$ since (1) and (2) consider only the area shaded by the yellow diagonal lines. Therefore, if there are ties, the AUC with (1) or (2) gives a different result from the actual AUC. For clarity, we give an example of AUC here, but the same problem occurs in the pAUC.

This is not a problem when using datasets with few ties in the score. However, this is a problem for tasks where there are many ties, such as network log

Table 1. Example of the network log where there is a tie.

Log	Destination IP address	URL	Port	Method	Status code	$\cdots$
Log1	192.0.2.1	http://www.cxample.com/RD.html	80	GET	200	$\cdots$
Log2	203.0.113.1	http://www.example.com/RD.html	80	GET	200	$\cdots$

analysis. As a simple example, consider the network log shown in Table 1 and assume that the scoring function is linear. Some well-known websites have multiple destination IP addresses for load distribution and operational efficiency, e.g., load balancer, even though the other payloads, e.g., URL and destination port number, are exactly the same. In this case, the features other than the destination IP address are exactly the same for Log1 and Log2. When converting some features such as the destination IP address into a feature vector, those are often vectorized with a one-hot vector since slightly different destination IP addresses, e.g., 192.0.2.1 and 192.0.2.2, are often completely different communication destinations even if they look similar. Therefore, if there is only one network log whose destination IP addresses are 192.0.2.1 and 203.0.113.1 in the training data, and both Log1 and Log2 are learned as benign logs, the scores of Log1 and Log2 are a tie since the weights, i.e., the coefficients of the features, given to the features 190.0.2.1 and 203.0.113.1 are equal.

4   PUMA

This section describes the details of PUMA. We focus on the pAUC in the low FPR interval since this interval is important for log analysis, but PUMA can be applied to any $[\alpha, \beta]$ interval. PUMA can maximize pAUC even if there are ties in the score of training data. PUMA increases the score of the malicious logs and lowers that of the benign logs in a suitable range of FPR.

4.1   Exact Definition of pAUC

In this subsection, we strictly define the pAUC considering ties for three intervals: $[\alpha, j_\alpha/n]$, $[j_\alpha/n, j_\beta/n]$, and $[j_\beta/n, \beta]$ interval. Figure 3 shows a part of the ROC curve when the score is a tie in the FPR interval including the $[\alpha, j_\alpha/n]$ interval, i.e., the diagonal line is a part of the ROC curve. $< 1 >\sim< 4 >$ is obtained from the definition of the TPR and FPR. $< 5 >$ and $< 6 >$ are obtained from $< 1 >\sim< 4 >$. When we calculate the area of the trapezoidal part corresponding to the pAUC in the three intervals, the exact pAUCs are as follows;

- $[\alpha, j_\alpha/n]$ interval (when $j_\alpha \neq 0$)

$$\frac{1}{mn(\beta - \alpha)}(j_\alpha - n\alpha)\left[\sum_{i=1}^{m} I(f(x_i^+; w) > f(x_{(j_\alpha)}^-; w)) - \sum_{j=1}^{n} I(f(x_j^-; w) > f(x_{(j_\alpha)}^-; w))\right.$$

$$\left.\cdot\frac{\sum_{i=1}^{m} I(f(x_i^+; w) = f(x_{(j_\alpha)}^-; w))}{\sum_{j=1}^{n} I(f(x_j^-; w) = f(x_{(j_\alpha)}^-; w))} + \frac{1}{2}(j_\alpha + n\alpha)\frac{\sum_{i=1}^{m} I(f(x_i^+; w) = f(x_{(j_\alpha)}^-; w))}{\sum_{j=1}^{n} I(f(x_j^-; w) = f(x_{(j_\alpha)}^-; w))}\right], (3)$$

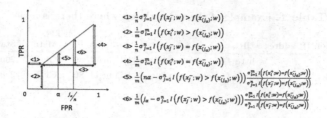

Fig. 3. A part of ROC curve with ties in the case of $[\alpha, j_\alpha/n]$ interval

– $[j_\alpha/n, j_\beta/n]$ interval

$$\frac{1}{mn(\beta - \alpha)} \sum_{i=1}^{m} \sum_{j=j_\alpha+1}^{j_\beta} \{I(f(\boldsymbol{x}_i^+; \boldsymbol{w}) > f(\boldsymbol{x}_{(j)}^-; \boldsymbol{w})) + \frac{1}{2}I(f(\boldsymbol{x}_i^+; \boldsymbol{w}) = f(\boldsymbol{x}_{(j)}^-; \boldsymbol{w}))\}, \quad (4)$$

– $[j_\beta/n, \beta]$ interval (when $j_\beta \neq n$)

$$\frac{1}{mn(\beta - \alpha)}(n\beta - j_\beta)\left[\sum_{i=1}^{m} I(f(\boldsymbol{x}_i^+; \boldsymbol{w}) > f(\boldsymbol{x}_{(j_\beta+1)}^-; \boldsymbol{w})) - \sum_{j=1}^{n} I(f(\boldsymbol{x}_j^-; \boldsymbol{w}) > f(\boldsymbol{x}_{(j_\beta+1)}^-; \boldsymbol{w}))\right.$$

$$\left. \cdot \frac{\sum_{i=1}^{m} I(f(\boldsymbol{x}_i^+; \boldsymbol{w}) = f(\boldsymbol{x}_{(j_\beta+1)}^-; \boldsymbol{w}))}{\sum_{j=1}^{n} I(f(\boldsymbol{x}_j^-; \boldsymbol{w}) = f(\boldsymbol{x}_{(j_\beta+1)}^-; \boldsymbol{w}))} + \frac{1}{2}(n\beta + j_\beta)\frac{\sum_{i=1}^{m} I(f(\boldsymbol{x}_i^+; \boldsymbol{w}) = f(\boldsymbol{x}_{(j_\beta+1)}^-; \boldsymbol{w}))}{\sum_{j=1}^{n} I(f(\boldsymbol{x}_j^-; \boldsymbol{w}) = f(\boldsymbol{x}_{(j_\beta+1)}^-; \boldsymbol{w}))}\right] (5)$$

The equations above are divided by $\beta - \alpha$ to normalize so that the maximum value is 1. When $j_\alpha = 0$, (3) becomes 0. When $j_\beta = n$, (5) becomes 0. Note that, in the case of $j_\alpha \neq 0$ or $j_\beta \neq n$, the denominator of the fractional part of (3) and (5) is not 0 but an integer greater than or equal to 1.

4.2   Formulation

This subsection describes the pAUC maximization method by using (3)–(5) as part of the objective function. First, since I is non-differentiable, we approximate the inequality part to the logistic sigmoid function σ and equal part to the exponential function ν whose maximum value is 1 as follows:

$$\sigma(\boldsymbol{x}_1, \boldsymbol{x}_2; \boldsymbol{w}) = [1 + \exp\{-(f(\boldsymbol{x}_1; \boldsymbol{w}) - f(\boldsymbol{x}_2; \boldsymbol{w}))\}]^{-1}, \quad (6)$$

$$\nu(\boldsymbol{x}_1, \boldsymbol{x}_2; \boldsymbol{w}) = \exp\left[-\frac{\{f(\boldsymbol{x}_1; \boldsymbol{w}) - f(\boldsymbol{x}_2; \boldsymbol{w})\}^2}{2\varsigma^2}\right], \quad (7)$$

where $\boldsymbol{x}_1$ and $\boldsymbol{x}_2$ are arbitrary vectors and ς is a hyperparameter that means variance. By applying these approximations, (3)–(5) can be rewritten as follows:

– $[\alpha, j_\alpha/n]$ interval (when $j_\alpha \neq 0$)

$$\frac{1}{mn(\beta - \alpha)}(j_\alpha - n\alpha)\left[\sum_{i=1}^{m} \sigma(\boldsymbol{x}_i^+, \boldsymbol{x}_{(j_\alpha)}^-; \boldsymbol{w}) - \sum_{j=1}^{n} \sigma(\boldsymbol{x}_j^-, \boldsymbol{x}_{(j_\alpha)}^-; \boldsymbol{w})\frac{\sum_{i=1}^{m} \nu(\boldsymbol{x}_i^+, \boldsymbol{x}_{(j_\alpha)}^-; \boldsymbol{w})}{\sum_{j=1}^{n} \nu(\boldsymbol{x}_j^-, \boldsymbol{x}_{(j_\alpha)}^-; \boldsymbol{w})}\right.$$

$$\left. + \frac{1}{2}(j_\alpha + n\alpha)\frac{\sum_{i=1}^{m} \nu(\boldsymbol{x}_i^+, \boldsymbol{x}_{(j_\alpha)}^-; \boldsymbol{w})}{\sum_{j=1}^{n} \nu(\boldsymbol{x}_j^-, \boldsymbol{x}_{(j_\alpha)}^-; \boldsymbol{w})}\right], \quad (8)$$

– $[j_\alpha/n, j_\beta/n]$ interval

$$\frac{1}{mn(\beta-\alpha)}\left[\sum_{i=1}^{m}\sum_{j=j_\alpha+1}^{j_\beta}\{\sigma(x_i^+, x_{(j)}^-; w) + \frac{1}{2}\nu(x_i^+, x_{(j)}^-; w)\}\right], \tag{9}$$

– $[j_\beta/n, \beta]$ interval (when $j_\beta \neq n$)

$$\frac{1}{mn(\beta-\alpha)}(n\beta - j_\beta)\left[\sum_{i=1}^{m}\sigma(x_i^+, x_{(j_\beta+1)}^-; w) - \sum_{j=1}^{n}\sigma(x_j^-, x_{(j_\beta+1)}^-; w)\frac{\sum_{i=1}^{m}\nu(x_i^+, x_{(j_\beta+1)}^-; w)}{\sum_{j=1}^{n}\nu(x_j^-, x_{(j_\beta+1)}^-; w)}\right.$$
$$\left. + \frac{1}{2}(n\beta + j_\beta)\frac{\sum_{i=1}^{m}\nu(x_i^+, x_{(j_\beta+1)}^-; w)}{\sum_{j=1}^{n}\nu(x_j^-, x_{(j_\beta+1)}^-; w)}\right]. \tag{10}$$

We introduce $s(x_i^+, x_{(j)}^-; w)$ that satisfies the following equations for each of $j = j_\alpha$, $j = J$ (J is an integer between $j_\alpha + 1$ and j_β), and $j = j_\beta + 1$ cases.

$$s(x_i^+, x_{(j_\alpha)}^-; w) = (j_\alpha - n\alpha) \cdot \left[\frac{1}{2}(j_\alpha + n\alpha)\frac{\nu(x_i^+, x_{(j_\alpha)}^-; w)}{\sum_{j=1}^{n}\nu(x_j^-, x_{(j_\alpha)}^-; w)}\right.$$
$$\left. + \sigma(x_i^+, x_{(j_\alpha)}^-; w) - \nu(x_i^+, x_{(j_\alpha)}^-; w)\frac{\sum_{j=1}^{n}\sigma(x_j^-, x_{(j_\alpha)}^-; w)}{\sum_{j=1}^{n}\nu(x_j^-, x_{(j_\alpha)}^-; w)}\right], \tag{11}$$

$$s(x_i^+, x_{(J)}^-; w) = \sigma(x_i^+, x_{(J)}^-; w) + \frac{1}{2}\nu(x_i^+, x_{(J)}^-; w), \tag{12}$$

$$s(x_i^+, x_{(j_\beta+1)}^-; w) = (n\beta - j_\beta) \cdot \left[\frac{1}{2}(n\beta + j_\beta)\frac{\nu(x_i^+, x_{(j_\beta+1)}^-; w)}{\sum_{j=1}^{n}\nu(x_j^-, x_{(j_\beta+1)}^-; w)}\right.$$
$$\left. + \sigma(x_i^+, x_{(j_\beta+1)}^-; w) - \nu(x_i^+, x_{(j_\beta+1)}^-; w)\frac{\sum_{j=1}^{n}\sigma(x_j^-, x_{(j_\beta+1)}^-; w)}{\sum_{j=1}^{n}\nu(x_j^-, x_{(j_\beta+1)}^-; w)}\right]. \tag{13}$$

Taking the logarithm of (8)–(10) and introducing a regularization term to mitigate the overfitting, the objective function E can be written as

$$E(w) = \log\left(\sum_{i=1}^{m}\sum_{j=j_\alpha}^{j_\beta+1} s(x_i^+, x_{(j)}^-; w)\right) - CR(w), \tag{14}$$

where $C \in \mathbb{R}$ is a hyperparameter and $R(w)$ is a regularization function. Common choices of $R(w)$ include L_1 and L_2 regularization and Elastic Net. The constant term was deleted since it does not affect the optimization calculation.

However, optimization calculation is difficult since (14) is not convex with respect to w. Following the expectation-maximization (EM) algorithm [14], we estimate w that provides a large $E(w)$ around the initial value by maximizing the lower bound of $E(w)$. Define $E_{LB}(w, q_{i(j)})$ as the lower bound of $E(w)$ that satisfies $E(w) \geq E_{LB}(w, q_{i(j)})$ where $q_{i(j)}$ holds $q_{i(j)} \geq 0$ and $\sum_{i=1}^{m}\sum_{j=j_\alpha}^{j_\beta+1} q_{i(j)} = 1$. Then, the following inequality holds from Jensen's inequality.

$$\log\left(\sum_{i=1}^{m}\sum_{j=j_\alpha}^{j_\beta+1} s(x_i^+, x_{(j)}^-; w)\right) \geq \sum_{i=1}^{m}\sum_{j=j_\alpha}^{j_\beta+1} q_{i(j)} \log s(x_i^+, x_{(j)}^-; w) - \sum_{i=1}^{m}\sum_{j=j_\alpha}^{j_\beta+1} q_{i(j)} \log q_{i(j)}. \tag{15}$$

Therefore, the lower bound of (14) becomes as follows:

$$E_{LB} = \sum_{i=1}^{m} \sum_{j=j_\alpha}^{j_\beta+1} q_{i(j)} \log s(\boldsymbol{x}_i^+, \boldsymbol{x}_{(j)}^-; \boldsymbol{w}) - \sum_{i=1}^{m} \sum_{j=j_\alpha}^{j_\beta+1} q_{i(j)} \log q_{i(j)} - CR(\boldsymbol{w}). \quad (16)$$

To find $\boldsymbol{w}$ that maximizes (16), $q_{i(j)}$ and $\boldsymbol{w}$ are optimized alternately. Let $\hat{\boldsymbol{w}}$ be a solution of the previous step. Using the method of a Lagrange multiplier under the constraint of $\sum_{ij} q_{i(j)} = 1$, we can derive $q_{i(j)}$ that maximizes E_{LB} as

$$\hat{q}_{i(j)} = \frac{s(\boldsymbol{x}_i^+, \boldsymbol{x}_{(j)}^-; \hat{\boldsymbol{w}})}{\sum_{i'=1}^{m} \sum_{j'=j_\alpha}^{j_\beta+1} s(\boldsymbol{x}_{i'}^+, \boldsymbol{x}_{(j')}^-; \hat{\boldsymbol{w}})}. \quad (17)$$

From the above, we can obtain the optimal parameter vector $\hat{\boldsymbol{w}}$ that maximizes the pAUC through maximizing the objective function (16) with respect to $\hat{\boldsymbol{w}}$ by using $\hat{q}_{i(j)}$ shown in (17). Note that (16) is affected only by benign logs within $j = j_\alpha \sim j_\beta + 1$, i.e., the benign logs out of the range do not matter.

4.3   Training

In summary, the algorithm of PUMA in the training process is as follows:

Algorithm 1. PUMA Algorithm

Require: Training datasets S^+ and S^-, scoring function $f(\boldsymbol{x}; \hat{\boldsymbol{w}})$, FPR interval α and
  β, regularization function $R(\hat{\boldsymbol{w}})$, and hyperparameter C.
Ensure: $\hat{\boldsymbol{w}}$.
 1: Initialize $\hat{\boldsymbol{w}}$.
 2: **while** not converged **do**
 3:     Update $\hat{q}_{i(j)}$ by calculating (17).
 4:     Update $\hat{\boldsymbol{w}}$ by maximizing (16).
 5:     Sort benign data points based on $\hat{\boldsymbol{w}}$ and f so that scores are in descending order.
 6: **end while**

Note that the reason PUMA requires iteration is that the benign data points need to be reordered for each iteration since the objective function changes if $\boldsymbol{w}$ is changed. Note that PUMA increases the time to create a classifier by $\mathcal{O}(N \log N)$ (N: number of data) compared with the baseline supervised learning method in the training. On the other hand, during the test, i.e., detection phase, it can be operated in the same order of time as the original supervised learning method.

Intuitively, the effect of considering ties is that PUMA lowers the scores of the benign logs that do not have the same score as the malicious logs than the scores of the benign logs that have the same score as the malicious logs.

Table 2. Number of hosts and network logs of datasets

	Dataset	Benign		Malicious		Acquisition date
		& of hosts	& of logs	& of malware samples	& of logs	
Test 1	Training	1,094	48,869	218	5,000	July 2017
	Validation	1,054	46,638	274	5,000	Aug. 2017
	Test	1,006	44,562	273	5,000	Sept. 2017
Test 2	Training	12,203	500,000	63	500	Feb. $\sim$ July 2017
	Validation	1,156	50,000	50	50	Aug. 2017
	Test	1,151	50,000	50	50	Sept. 2017

Scoring Function. The scoring function $f(x; w)$ plays an important role in maximizing the pAUC. Ideally, the scoring function should satisfy $f(x^+; w) > f(x^-; w)$ for any x^+ and x^-. In this paper, we consider the linear scoring function of the form $f(x; w) = w^T x$ since it is desirable in log analysis. When a suspicious log is detected in actual operation, the network operators manually analyze details such as whether a log that is classified as malicious is really malicious and whether a countermeasure needs to be taken such as blocking communication. Therefore, the reasons for detection should be made clear to save time. Using a linear function as the scoring function, we can understand which feature is effective by investigating w. Note that PUMA is applicable regardless of whether the scoring function is linear or nonlinear.

Optimization Method. We adopted L_2 regularization to prevent overfitting and used the L-BFGS algorithm [15], a kind of quasi-Newton methods, to maximize (16) since it is used in various machine learning methods. Note that the L-BFGS algorithm does not guarantee a global optimal but a local optimal. Therefore, we should choose the initial value carefully. In this paper, we used w obtained by LR as the initial value. Note that an optimization algorithm used to maximize (16) is not limited to the L-BFGS algorithm. Other gradient methods, e.g., SGD algorithm, can be used. We also considered this algorithm to be converged when $||1 - E_{new}/E_{old}|| \leq \epsilon$ holds, where E_{new} is E_{LB} after update and E_{old} is E_{LB} before update. ϵ is a very small value, e.g., 0.001, and that we set the maximum number of iterations in consideration of the possibility of non-convergence.

5   Datasets

We collected proxy logs from an actual large enterprise network located in Japan. Proxy logs include source/destination IP address, source/destination port number, number of source/destination packets, number of request/response bytes, URL, user agent, HTTP status code, HTTP method, and timestamp. Proxy logs are widely used for log analysis to detect malicious logs since a large amount of malware uses HTTP as transport in drive-by-download and communications between C&C servers. We created the benign datasets using logs when no incidents were reported by operators and malware was not detected from multiple commercial tools installed at different monitoring points from gateway to end-point. Through this process, we avoided mixing malicious logs into benign logs.

For the malicious datasets, we collected malware samples from VirusTotal [16] and obtained HTTP communication logs through dynamic analysis in a sandbox [17]. All malware samples were found to have different SHA1 when they were first seen and diversified with different types of malware family. According to our investigation, 73 malware families are included in 4,941 malware samples (126,171 logs). We examined the number of malware families with ESET [18]. We excluded logs whose destination domain names are listed in the Alexa top 1 million [19] to avoid mixing benign logs into malicious logs.

We prepared two patterns of datasets, i.e., Tests 1 and 2. Table 2 shows the number of hosts and network logs of the training, validation, and test datasets of Tests 1 and 2. The rate of benign to malicious logs of Test 2 was about 1,000 to 1 to simulate an imbalanced dataset assumed in actual operation. Note that duplicated logs, i.e., logs from one host to the same URL, were eliminated to improve classification performance.

6   Evaluation

6.1   Experimental Conditions

We considered a binary classification problem of whether a proxy log is benign or malicious. We compared PUMA with previous pAUC maximization and conventional supervised learning methods to evaluate the effectiveness of PUMA.

We used the $pAUC_{[\alpha,\beta]}$, AUC, and $TPR_{FPR=X}$ as performance measures. The $pAUC_{[\alpha,\beta]}$ represents the pAUC between α and β, i.e., the $pAUC_{[\alpha,\beta]}$ is calculated by substituting α and β into (2). We define the $TPR_{FPR=X}$ as the TPR when adjusting the threshold so that the FPR = X. The $TPR_{FPR=\alpha}$ and $TPR_{FPR=\beta}$ are shown in Fig. 1 for reference. Note that the $TPR_{FPR=X}$ and $pAUC_{[0,X]}$ generally take different values.

We adjusted the hyperparameters to maximize the designated pAUC or AUC with hold-out validation because we needed to consider time series. Specifically, if we use cross-validation, it is likely to create a classifier by learning malware that was found later in a time series and detect malware that was found earlier, which is not desirable. Therefore, we collected training, validation, and test datasets in chronological order from the oldest acquisition date, then carried out parameter tuning with the validation dataset, and tested with the test dataset.

6.2   Feature Extraction

We used features often used in many studies [20–22] for fair evaluation. Table 3 lists both word-based and statistical-based features. Using only statistical-based features does not result in high classification performance. On the other hand, using word-based features in addition to statistical-based features tends to result in better performance. Bartos's results [20] also suggest that using only statistical-based features does not result in high classification performance, e.g., the AUC was less than 0.7. Therefore, we prepared many word-based features.

Note that PathElement in Table 3 means the element of each word separated by "/" in a URL path. For example, consider the URL "http://www.example.com/RD/index.php." In this case, the PathElements are "RD" and "index.php." The AS number, country, and city were resolved with the MaxMind GeoIP Lite database [23]. We also replaced the UserAgent value with "well-known" for well-known browsers, which was determined using Python UA parser [24].

Table 3. Features for actual proxy logs

Word-based features	Features	Statistical-based features	Features
HTTP-based	FQDN	Length	URL
	TLD		Domain
	PathElement		FQDN
	QueryKey		Path
	QueryParameters		Query
	User Agent		Filename
	Method		Extension
TCP/IP-based	Destination IP address	Presence	Symbols in Path*
	Destination port numer		Extension in URL*
Enrichment	AS number	& of Numbers	URL
	Country		FQDN
	City		Path

To create word-based features, we used a bag-of-words model [25]. We regarded all patterns (e.g., 192.0.2.5, 198.51.100.5, and 203.0.113.5) appearing in each feature (e.g., destination IP address) as one "element" and assigned 1 or 0 depending on whether the "element" appeared in a log. Statistical-based features were normalized by dividing by roughly the largest value assumed in the training data. The length of a part or all of a URL is normalized by dividing it by 2083, which is the maximum number of characters in a URL. Feature vectors marked with * were normalized by assigning present = 1 or absent = 0.

6.3 Methods Evaluated

This subsection shows the machine learning methods used for evaluation. We mainly used linear machine learning methods since we can understand which feature contributes to the classification results, which is desirable in log analysis.

PUMA. We applied PUMA with L_2 regularization, i.e., $R(\boldsymbol{w}) = ||\boldsymbol{w}||^2$, and determined a penalty parameter $C \in \{10^A\}_{A=-10}^{10}$ and variance $\varsigma \in \{10^A\}_{A=-5}^{-1}$. At this time, we set α and β to the same values used in the performance measures, $\epsilon = 0.001$ and the maximum number of iterations as 100.

PUMA-Notie. We also implemented PUMA *without* considering ties, hereinafter called PUMA-notie, for reference. When we apply (6) to (2) to make it differentiable, the following equation is obtained.

$$\frac{1}{mn(\beta - \alpha)} \sum_{i=1}^{m} \Big[(j_\alpha - n\alpha) \cdot \sigma(\boldsymbol{x}_i^+, \boldsymbol{x}_{(j_\alpha)}^-; \boldsymbol{w})$$

$$+ \sum_{j=j_\alpha+1}^{j_\beta} \sigma(\boldsymbol{x}_i^+, \boldsymbol{x}_{(j)}^-; \boldsymbol{w}) + (n\beta - j_\beta) \cdot \sigma(\boldsymbol{x}_i^+, \boldsymbol{x}_{(j_\beta+1)}^-; \boldsymbol{w})\Big]. \tag{18}$$

If we replace (8)–(10) with the above equation, (11) and subsequent equations are similarly obtained. Hyperparameters are selected from exactly the same candidates as PUMA.

Table 4. Performance measures with proxy logs

	Method	pAUC$_{[0,0.001]}$	pAUC$_{[0,0.01]}$	AUC	TPR$_{FPR=0.001}$	TPR$_{FPR=0.01}$
Test A	PUMA	**0.5792**	**0.7781**	**0.9939**	**0.6706**	**0.8702**
	PUMA-notie	0.5361	0.7592	0.9934	0.6400	0.8617
	SVMpAUC	0.0650	0.2891	0.9703	0.2356	0.4057
	LR	0.5063	0.7545	0.9932	0.5890	0.8508
	SVM	0.5033	0.7587	0.9917	0.6115	0.8607
	MLP	0.5185	0.7584	0.9934	0.6322	0.8572
Test B	PUMA	**0.7620**	**0.8633**	**0.9947**	**0.82**	**0.94**
	PUMA-notie	0.7384	0.8478	0.9946	0.78	0.90
	SVMpAUC	0.7101	0.8246	0.9898	0.72	0.86
	LR	0.6797	0.8389	0.9934	0.74	0.90
	SVM	0.7327	0.8312	0.9891	0.76	0.88
	MLP	0.7334	0.8389	0.9940	0.76	0.90

SVMpAUC. We implemented a previous pAUC maximization method [12,26], hereinafter called SVMpAUC. We chose $C \in \{10^A\}_{A=-10}^{10}$ and $\epsilon \in \{10^A\}_{A=-5}^{-1}$. The symbols here have the same roles as in PUMA (please see [12] for details). Note that SVMpAUC assumes that there are no ties.

LR. We used L$_2$ regularization and chose penalty parameters from $\{10^A\}_{A=-10}^{10}$.

SVM. We used linear kernel and L$_2$ regularization and chose penalty parameters from $\{10^A\}_{A=-10}^{10}$. Note that the rbf kernel is not suitable since it is difficult to understand which features are effective for classification performance.

MLP. We implemented multilayer perceptron (MLP) for reference. The settings are as follows. Number of hidden layers: 3, number of neurons: $\{10^A\}_{A=2}^{4}$, activation function: relu, solver: adam, penalty parameter: $\{10^A\}_{A=-10}^{10}$, learning rate: $\{10^A\}_{A=-10}^{-1}$, and numerical stability in adam: 10^{-8}. Note that MLP is a nonlinear machine learning method which is not suitable for log analysis.

6.4 Experimental Results

Table 4 lists the results of performance measures. We set $\alpha = 0$ and $\beta = 0.01$ or 0.001. Note that these value should be determined from the FPR that can be tolerated from the operating status of network operators in actual operation.

PUMA performed the best for all performance measures. PUMA and PUMA-notie outperformed LR, SVM, and MLP since PUMA and PUMA-notie maximize the pAUC. In addition, PUMA outperformed PUMA-notie and SVMpAUC since PUMA considers ties. SVMpAUC did not perform well since there were many ties. In the training data used in the Tests 1 and 2, the number of logs for which the score is a tie was 12,322 (25%) and 117,245 (23%) for benign training data and 967 (19%) and 81 (16%) for malicious training data, respectively.

Please pay particular attention to the results of $\text{TPR}_{\text{FPR}=0.001}$ and $\text{TPR}_{\text{FPR}=0.01}$ since the purpose of this paper is to maximize the TPR in a low FPR interval. Note that the performance improvement by PUMA is large, though it seems there is not much difference at first glance. For example, the difference between $\text{TPR}_{\text{FPR}=0.001}$ with PUMA and SVM in Test A were $0.6706 - 0.6115 = 0.0591$. This means that if you operate with FPR = 0.001, i.e., it allows that the classifier misclassifies $44,562 \times 0.001 \simeq 44.6$ benign logs as malicious, you can detect extra $5,000 \times 0.0591 \simeq 295.5$ malicious logs. This is a big difference since network operators want to detect suspicious logs without overlooking.

Table 5. Performance measures of related works

Works	$\text{TPR}_{\text{FPR}=x}$	AUC	Accuracy	TPR	FPR	Precision	F-measure
Mirsky et al. [3]	✓	✓		✓			
Copty et al. [4]	✓	✓		✓	✓		✓
Wressnegger et al. [5]	✓			✓	✓		
Hendler et al. [6]	✓	✓		✓	✓		
Pereira et al. [7]				✓	✓	✓	
Zhang et al. [8]		✓		✓	✓		✓
Stergiopoulos et al. [9]			✓	✓		✓	✓

Note that the effectiveness of PUMA was not limited to this evaluation using proxy logs. Although details are not shown here due to space considerations, the effectiveness of PUMA was also confirmed when we used public datasets, NSL-KDD [27] and ISCX-URL2016 [28].

7 Related Work

Most studies using machine learning methods for malicious log detection have used various performance measures to evaluate the performance of a classifier, including the $TPR_{FPR=x}$, AUC, accuracy, TPR, FPR, precision, and F-measure. To the best of our knowledge, classification performance has not been evaluated by using the pAUC. Table 5 lists the performance measures used in state-of-the-art studies to detect malicious logs. Researchers usually evaluate the performance of their machine learning methods by focusing on the $TPR_{FPR=x}$ [3–6]. However, these studies did not directly maximize the pAUC.

Several machine learning studies [29,30] have focused on the AUC as a performance measure for binary classification and proposed methods for directly optimizing the AUC, though most conventional supervised learning methods are trained to maximize accuracy, i.e., to minimize the error rate. In fact, well-known machine learning methods in many books [14] and machine learning libraries, e.g., scikit-learn [31], use an objective function that maximizes accuracy.

Although the number of previous studies on pAUC maximization methods is limited, few studies [11–13] have focused on the pAUC and proposed optimization methods for directly maximizing the pAUC as a more sophisticated strategy. Dodd et al. [10] proposed a pAUC maximization method by using (2) as an objective function. However, the calculation cost is high since (2) is a non-convex nonlinear objective function. Narasimhan et al. [11,12] proposed a training method for directly maximizing the pAUC only in an SVM with a linear kernel case. To reduce the calculation cost of optimizing (2), they also proposed an optimized calculation method using SVMstruct [12] or imposing constraints on the upper bound [11]. Ueda et al. [13] showed how to directly maximize the pAUC with a nonlinear scoring function. We devised a formulation for considering ties in the scores of training data and facilitating the optimization calculation to make PUMA a suitable method for analyzing logs.

8 Conclusion

We proposed Partial area under the curve (AUC) Utilization for Malicious log Analysis (PUMA) for detecting malicious activities. PUMA directly maximizes the true positive rate (TPR) under the constrained false positive rate (FPR) interval, i.e., partial AUC (pAUC). Since malicious logs need to be detected with an extremely low FPR to reduce operational burden, PUMA is a practical supervised learning method for log analysis. The advantage of PUMA over existing pAUC maximization methods is that it can successfully maximize pAUC even if there is a tie in the score of training data scored by the classifier. In the log analysis, scores in the training data tend to be a tie since network logs tend to have many similar feature vectors. We compared PUMA with conventional pAUC maximization and supervised learning methods and demonstrated that it outperformed them by using proxy logs from a large enterprise network.

References

1. Griffin, K., Schneider, S., Hu, X., Chiueh, T.: Automatic generation of string signatures for malware detection. In: Kirda, E., Jha, S., Balzarotti, D. (eds.) RAID 2009. LNCS, vol. 5758, pp. 101–120. Springer, Heidelberg (2009). https://doi.org/10.1007/978-3-642-04342-0_6

2. Jang, J., Brumley, D., Venkataraman, S.: BitShred: feature hashing malware for scalable triage and semantic analysis. In: CCS, pp. 309–320. ACM (2011)

3. Mirsky, Y., Doitshman, T., Elovici, Y., Shabtai, A.: Kitsune: an ensemble of autoencoders for online network intrusion detection. In: NDSS. Internet Society (2019)

4. Copty, F., Danos, M., Edelstein, O., Eisner, C., Murik, D., Zeltser, B.: Accurate malware detection by extreme abstraction. In: ACSAC, pp. 101–111. ACM (2018)

5. Wressnegger, C., Yamaguchi, F., Arp, D., Rieck, K.: Comprehensive analysis and detection of flash-based malware. In: Caballero, J., Zurutuza, U., Rodríguez, R.J. (eds.) DIMVA 2016. LNCS, vol. 9721, pp. 101–121. Springer, Cham (2016). https://doi.org/10.1007/978-3-319-40667-1_6

6. Hendler, D., Kels, S., Rubin, A.: Detecting malicious powershell commands using deep neural networks. In: AsiaCCS, pp. 187–197. ACM (2018)

7. Pereira, M., Coleman, S., Yu, B., DeCock, M., Nascimento, A.: Dictionary extraction and detection of algorithmically generated domain names in passive DNS traffic. In: Bailey, M., Holz, T., Stamatogiannakis, M., Ioannidis, S. (eds.) RAID 2018. LNCS, vol. 11050, pp. 295–314. Springer, Cham (2018). https://doi.org/10.1007/978-3-030-00470-5_14

8. Zhang, J., Jang, J., Gu, G., Stoecklin, M.P., Hu, X.: Error-sensor: mining information from HTTP error traffic for malware intelligence. In: Bailey, M., Holz, T., Stamatogiannakis, M., Ioannidis, S. (eds.) RAID 2018. LNCS, vol. 11050, pp. 467–489. Springer, Cham (2018). https://doi.org/10.1007/978-3-030-00470-5_22

9. Stergiopoulos, G., Talavari, A., Bitsikas, E., Gritzalis, D.: Automatic detection of various malicious traffic using side channel features on TCP packets. In: Lopez, J., Zhou, J., Soriano, M. (eds.) ESORICS 2018. LNCS, vol. 11098, pp. 346–362. Springer, Cham (2018). https://doi.org/10.1007/978-3-319-99073-6_17

10. Dodd, L., Pepe, M.: Partial AUC estimation and regression. Biometrics **59**(3), 614–623 (2003)

11. Narasimhan, H., Agarwal, S.: SVMpAUCtight: a new support vector method for optimizing partial AUC based on a tight convex upper bound. In: KDD, pp. 167–175. ACM (2013)

12. Narasimhan, H., Agarwal, S.: A structural SVM based approach for optimizing partial AUC. In: ICML, vol. 28. (2013)

13. Ueda, N., Fujino, A.: Partial auc maximization via nonlinear scoring functions (2018). arXiv preprint arXiv:1806.04838v1

14. Bishop, C.: Pattern Recognition and Machine Learning. Springer-Verlag, Heidelberg (2006). https://doi.org/10.1007/978-1-4615-7566-5

15. Zhu, C., Byrd, R., Lu, P., Nocedal, J.: Algorithm 778: l-bfgs-b: fortran subroutines for large-scale bound-constrained optimization. ACM TOMS **23**(4), 550–560 (1997)

16. Virustotal. https://www.virustotal.com/, Accessed 6 Sept 2021

17. Anonymous, our organization's paper

18. ESET Homepage. https://www.eset.com/, Accessed 6 Sept 2021

19. AWS. http://s3.amazonaws.com/alexa-static/top-1m.csv.zip, Accessed 6 Sept 2021

20. Bartos, K., Sofka, M.: Optimized invariant representation of network traffic for detecting unseen malware variants. In: USENIX, pp. 807–822. USENIX (2016)
21. Nelms, T., Perdisci, R., Ahamad, M.: ExecScent: mining for new c&c domains in live networks with adaptive control protocol templates. In: USENIX, pp. 589–604. USENIX (2013)
22. Ma, J., Saul, L., Savage, S., Voelker, G.: Learning to detect malicious urls. ACM TIST **2**(3), 30:1-30:24 (2011)
23. Python. https://pypi.org/project/geoip2/, Accessed 6 Sept 2021
24. Python. https://pypi.python.org/pypi/ua-parser, Accessed 6 Sept 2021
25. Zhang, Y., Jin, R., Zhou, Z.: Understanding bag-of-words model: a statistical framework. Int. J. Mach. Learn. Cybern. **1**(4), 43–52 (2010). https://doi.org/10.1007/s13042-010-0001-0
26. Narasimhan, H.: http://clweb.csa.iisc.ac.in/harikrishna/Papers/SVMpAUC/index.html, Accessed 6 Sept 2021
27. Tavallaee, M., Bagheri, E., Lu, W., Ghorbani, A.: A detailed analysis of the KDD cup 99 data set. In: CISDA, pp. 1–6. IEEE (2009)
28. UNB. https://www.unb.ca/cic/datasets/url-2016.html, Accessed 6 Sept 2021
29. Herschtal, A., Raskutti, B.: Optimising area under the roc curve using gradient descent. In: ICML. ACM (2004)
30. Calders, T., Jaroszewicz, S.: Efficient AUC optimization for classification. In: Kok, J.N., Koronacki, J., Lopez de Mantaras, R., Matwin, S., Mladenič, D., Skowron, A. (eds.) PKDD 2007. LNCS (LNAI), vol. 4702, pp. 42–53. Springer, Heidelberg (2007). https://doi.org/10.1007/978-3-540-74976-9_8
31. Scikit-learn. https://scikit-learn.org/stable/, Accessed 6 Sept 2021

A Drift Aware Hierarchical Test Based Approach for Combating Social Spammers in Online Social Networks

Darshika Koggalahewa[1]([envelope]), Yue Xu[1], and Ernest Foo[2]

[1] Queensland University of Technology, Brisbane, QLD, Australia
darshikaniranjan.koggalahewa@hdr.qut.edu.au, yue.xu@qut.edu.au
[2] Griffith University, Brisbane, QLD, Australia
e.foo@griffith.edu.au

Abstract. Spam detection in online social networks (OSNs) have become an immensely challenging task with the nature and the use of online social networks. The spammers tend to change their behaviors over time which is commonly known as the spam drift in Online Social networks. The most popular spam combatting approaches, such as classification will grew mere unsuccessful when the drift is present since old labels cannot be used to train the new classifiers. This paper presents the comprehensive results of an approach developed to identify the drift by using a set of datasets. It introduces a set of real time user interest-based features which can be used for spam drift detection in OSNs. The paper presents a hierarchical test-based drift detection approach for spam drift learning overtime, where it learns features in an unsupervised manner. The feature similarity differences, KL divergence and the Peer Acceptability of a user is then used to detect and validate the drifted content of spam users in the real time. The system automatically updates the learning model and relabeled the users with new predicted label for classification with the new knowledge acquired through drift detection. The results are encouraging. The error rate of the classifiers with drift detection is below 1%.

Keywords: Spam detection · Concept drift · Hierarchical test

1 Introduction

Spam detection is a challenging process where spammers are trying to disguise themselves as genuine users to bypass spam detection systems. Online Social Networks (OSNs) are highly vulnerable to spamming activities due to their diversified nature and activities. OSNs have used different types of machine learning approaches such as classifiers and non-machine learning approaches such as blacklisting to combat spammers. However, machine learning based methods suffer from the problem of "spam drifting" where, classifiers trained using the old data cannot be used to recognize new spammers in OSNs. This occurs due to the change of distribution between predicted and original variable over time. This is known as concept drift.

© Springer Nature Singapore Pte Ltd. 2021
Y. Xu et al. (Eds.): AusDM 2021, CCIS 1504, pp. 47–61, 2021.
https://doi.org/10.1007/978-981-16-8531-6_4

Concept drift is the problem of changing the properties of target variables of a predicted model overtime in an unexpected manner [1]. The properties of previous data are not relevant for the new data. This problem has been identified and explored by different researchers over the last two decades [2–4, 8]. Nevertheless, the problems of "how to accurately detect concept drift in unstructured and noisy datasets, how to quantitatively understand concept drift in an explainable way, and how to effectively react to drift by adapting related knowledge" remain as challenging problems [5]. Machine learning approaches used for spam detection in OSNs use different types of features. They can be categorized in to four main types, content-based, account-based, graph based and campaign-based features [6]. Content based features were developed from both statistical measurements such as number of hash tags, number of mentions as well as the actual tweet content. The classifiers used for OSN spam detection used a combination of these feature groups. Applicability of traditional concept drift techniques in twitter spam drifting is challenging due to multiple reasons: (1) Character limitation in twitter (280 characters maximum), (2) use of combined features from multiple groups, (3) the noise available in twitter posts, (4) platform specific language usage patterns, (5) use of emojis and hashtags to communicate some messages and (6) complexity of features [5, 7, 14]. These limitations and the open issues motivated this paper to propose a spam drift detection and adaptation approach unique for twitter OSNs.

Drift detection in short text is challenging compared to the drift detection in long text. Short text does not contain sufficient statistical measures to make text analysis meaningful. The limited context of short text makes the change detection difficult. The data stream would be continuous and open ended with high volume and variation. All these three limitations make drift learning in short text more difficult.

Literature suggests that many of the spam drifting approaches proposed to deal with generic concept drifting types, apply traditional detection and treatment approaches. Traditional spam drifting models validate the performance of the trained model over a labelled dataset and drift is detected when performance drops. Since such methods were time consuming, a set of automated methods were introduced to detect concept drift by monitoring the feature distribution to detect the drift. The problem with such systems is that they are fundamentally prone to false alarms and cannot distinguish between changes that affect a classifier's performance [6–8]. Nevertheless, it is essential to consider both generic concept drift types as well as domain specific drifting methods such as Adversarial Drifts mentioned earlier for a successfully handling the spam drift.

The paper presents and approach to learn (detect and adapt) concept drift in Twitter OSN. First, we develop a hierarchical drift detection approach for concept drift detection in Twitter OSN users. Hierarchical detection algorithms have two layers in their approach, detection, and validation. The drifted users are detected as the output of our approach. Since OSN spammers have different types of features, our detection was operated at two levels. First, we considered the users information interest derived through user posts to detect the drift. Then we used a set of spammer features (both content and non-content based) to detect the drift. For the validation of the algorithm, we use our previous work in peer acceptance-based spam detection [19]. We considered the two drift types of sudden drift and gradual drift with adversarial drift in spam users. Drift detection is followed by a drift adaptation approach where we update the model with drifted features and new labels.

The research is evaluated with two datasets, called Peer Acceptance Data Collection (PADC) and User Information Interest dataset (UIID). The experiment results showed that the proposed drift learning can deal with twitter spam drifting by reducing the classifier error rate under 1%. There were set of state-of-the-art systems used for the experiments in both drifting behavior investigation and handling spam drifting. The contribution of the paper can be summarized as follows.

- Development of an unsupervised hierarchical drift detection approach by considering both generic concept drift types as well as the specific concept drift, "Adversarial Drifts"
- An unsupervised approach for drift adaptation in Twitter OSN

Section 2 of the research paper describes the current contributions in spam drift detection and adaptation. Section 3 describes the Drift detection approach and the approach introduced for drift handling introduced in the research. Section 4 explains the experimental results along with the observations obtained from the research. A discussion on strengths and weaknesses is described in Sect. 5. Section 6 concludes the research work.

2 Background and Related Work

Concept drift is the phenomena of "relation between the input data and the target variable changes over time" [15]. The concept drift is common in most of our real-world scenarios. The areas such as, object detection [24], decision making [25], machine learning and robotics, Artificial intelligence [26, 27]. The models trained using older data may not be significant with drifting. There were different efforts to handle concept drift in real world scenarios. Research conducted for concept drift detection can be classified in to four main types: Error rate-based drift detection, Data Distribution-based Drift Detection, Multiple Hypothesis test-based drift detection [5]. Error rate-based algorithms are designed to track error rates of classifiers. The most common error rate-based algorithm is the "Drift Detection Method (DDM)" [17]. It defines a warning level and a drift level for its detection. There are many systems adapted the error-based learning algorithm which improved the limitations in DDM. Some of them are Learning with Local Drift Detection (LLDD) [18]. Heoffding's inequality-based Drift Detection Method (HDDM) [20]. Data distribution-based drift detection algorithms, use a function to detect the dissimilarity between two distributions current and historical. They used multiple time windows and calculated the difference between two time windows [21, 22]. Multiple hypothesis-based drift detection algorithms use hierarchical layers with multiple schemas. It is an emerging technology where, Hierarchical Change-Detection Tests was the first to introduce this new technique. They used one detection layer and a validation layer in their detection [23].

There are four main types of concept drift; sudden drift (A new concept occurs within a short time), gradual drift (A new concept gradually replaces an old one over a period), incremental drift (An old concept incrementally changes to a new concept over a period of time) and reoccurring concepts (An old concept may reoccur after some time) [5]. Sudden drifts and gradual drifts are most common in the OSN spammers. Some spammers display the behavior of reoccurring concepts, but most spammers' life span is comparably short. We considered sudden drift and gradual drift in our experiments. A complete learning process under concept drift consists of three main phases, drift detection, drift understanding and drift adaptation. Drift detection refers to the techniques and measurements designed to identify drift changes, while drift understanding finds evidence that the drift occurred (i.e.: when and where). Drift adaptation is the process of updating the existing learning model to address the impact of drifting.

Most of the existing work designed for drift detection and adaptation, handles this problem as a domain agnostic one. All the changes in data distributions were treated equally and causes of such changes are ignored. Though this equal treatment is good for handling generic concept drift, it is questionable for spam drifting. Spammers change their behaviors to evade detection. Spammers typically learn the nature of deployed classifier model, using different techniques and exploit this information to generate attacks [9, 10, 14]. Such types of drifts are named "Adversarial Drifts" and it is a sub type of concept drift. It is different from the traditional concept drift by "1) The drift is a result of changes to the malicious class samples only, 2) The drift is a function of the deployed classifier model, as the adversary reverse engineers and actively tries to evade it" [10].

The state of art in Twitter spam drift detection can be explained in three categories with reference to the classification features used in Twitter spam detection systems: content-based Account-based, graph-based and hybrid approaches. To represent data distribution, it needs to firstly extract data by using raw, statistical or machine learning-based approaches [8, 11]. Raw features are Twitter attributes that are directly extracted from a dataset and sometimes are sufficiently distinguishable to be directly applied for monitoring changes in decision boundaries in Twitter spam detection approaches. Statistical features are extracted by characterizing data through testing hypothesis, whereas machine learning-based features are extracted through neural network or deep learning layer structures.

An account-based approach puts emphasis on analyzing user, account and tweet content information. For example, user and account information includes name, number of follower and followees, account age, description and many other account information features. Name convention features are information regarding usernames, its account statistics and information, including, length of the names, ratio of name length, and similarity level in the names. Account information features cannot be changed once an account is created, including account creation time and verification status.

3 An Approach of Spam Drift Detection Based on User Behavior Changes

In this section we present a hybrid approach of spam drift detection based on a set of content, non-content-based and user behavior-based features. We identify the features from the datasets, analyze the changes of these features in terms of their distribution overtime. If the changes are significant, we update and predict a new label to handle the drift occurred in the user. We have considered a set of features which are commonly used in the recent literature and a set of specific features that we derived through our spam detection approach which detect spammers based on user interests.

3.1 Hierarchical Test-Based Approach for Drift Detection

There are different types of drift detection approaches, e.g., "Error rate-based drift detection, Data Distribution-based Drift Detection, Multiple Hypothesis Test Drift Detection" [5]. Among them multiple hierarchical hypothesis test-based drift detection algorithms can be considered as a new emerging technology. Hierarchical approaches have a detection layer and a validation layer. The detection layer consists of different detection algorithms each of which is a sub layer under the detection layer. The validation layer consists of a scheme for validation. Most of the existing hierarchical detection layers use only one detection algorithm. The detection layer presented in this paper consists of three detection algorithms, (1) user's information interest-based detection algorithm, (2) algorithm of spam detection based on Kullback-Leibler (KL) Divergence [12], or the relative entropy of the distribution and (3) algorithm of similarity between non-content spam feature spaces. The validation layer consists of a validation scheme for drifted users based on our previous work on peer acceptance-based spam detection. Three algorithms provide a set of drifted users. Finally, we use a validation layer to validate the drifted users detected in previous algorithms. When the drifted user set is confirmed, we update classifier model with drifted users with their features. The new predicted label for the user is used to train the classifier in new window. Sections 3.2, 3.3 and 3.4 describe the methodology applied in each detection layer. Figure 1 depicts the architecture of the hierarchical drift detection approach.

As depicted in Fig. 1, the input consists of two adjacent time windows, W_i and W_{i+1}. W_i refers to a time period of one month containing users' post content. W_{i+1} refers to the next month of the data stream. For each time window, we have the users' post content from which the feature set described in Sect. 3.3 and the user profile of each user are generated. Layer one of our approaches use user's interest-based features to detect the drifted users in new time window W_{i+1} Instead of one detection layer we proposed several algorithms for drift detection. Features used in layer 1 are described in Sect. 3.1. We send the drifted users detected in layer 1 through layer 2. Layer two is based on KL divergence. In addition to the detection in two layers, we used a common set of features for spam detection in layer 3.

3.1.1 User Information Interest for Drift Detection

The Algorithm 1 in detection layer 1 is to detect drift based on user's information interest distribution over a certain topic. First, we analyzed the user's information interest over a certain topic. Analysis showed that, for a drifted user the content similarity between two adjacent time windows will be changed since they try to modify the content compared to the previous content. They try to modify their spam strategy and include some relevant content for the topic in attempting to avoid being detected by similarity-based detectors. Hence their content similarity of a topic between two consecutive time windows will be changed overtime. They also show a similarity difference between their own topic

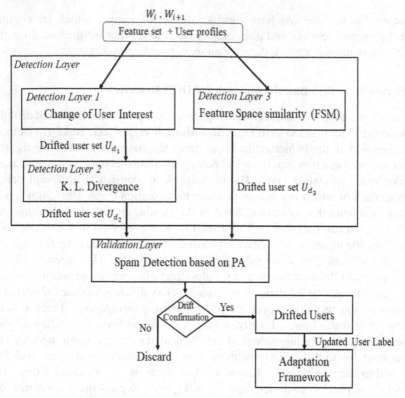

Fig. 1. Architecture of hierarchical testing-based drift detection

content and the overall content for the topic. Their topic usage could show a sudden change as well. The topics used in their posts have a sudden change for drifted users while ordinary users gradually change their topics. We made the following hypothesis to develop the drift detection in layer 1.

H_1: Drifted users should have a higher similarity for the same topic in two adjacent time windows.

H_2: The similarity difference between an individual user's information interest and Overall topic interest should be reduced in two adjacent time windows.

H_3: If a user frequently changes their topics over multiple time windows and a higher percentage of these topics represent the set of frequent topics from that time window, he is potentially a drifted user.

We used users' posts to derive the user's information interest. User content profile $\mathcal{CI}$ is a tensor, $\mathcal{CI} \in \mathbb{N}^{|U| \times |T| \times |W|}$ where each user's content profile is represented as, $\mathcal{CI}(u_x, t_k, w_n)$. In there, t_k is the topic and $\mathcal{CI}(u_x, t_k, w_n)$ is the tf-idf value of w_n for user, u_x posts in topic t_k. We use, vector $\overrightarrow{\mathcal{CI}_w}(u_x, t_k) \in \mathbb{N}^{|W|}$ to represent user, u_x 's information interest in window w. The topic content, T_t represents the average interest from all users who used topic t. It is denoted as $\vec{T}_t = \frac{1}{m} \sum_{u_x \in U} \vec{CI}(u_x, t)$ where x = 1, 2,…,m. Users use many topics in their posts. But they may not interest in all the topics. To determine a set of interested topics, we calculate the similarity between user's content vector $\vec{CI}$ and the Topic content vector, $\vec{T}_t$. User's interested topics are denoted as, $UT(u)$. Algorithm 1 is generated based on the above hypothesis and the definitions to identify drifted users. W_i and W_{i+1j} are two adjacent time windows with the length of one month as defined in Sect. 3.1.

Algorithm 1: Drifted user detection based on User Information Interest

---

Input: Data for two time windows W_i, W_{i+1}

    User set U

      User content profile CI,

      Topic content representation T_j, $j = 1, … n$,

      Representative topics for each user $UT(u_i)$ $i = 1, …, m$

      experimental thresholds k, l

Output: Drifted User set U_d

---

1.   Set $U_{d_1} = \{\}$, Drifted $V_1 = False$, Drifted $V_2 = False$, Drifted $V_3 = False$,

2.   For each user u_a in U in two adjacent time windows W_i, W_{i+1}

3.   For each topic $t_k \in UT(u_a)$

4.   if $sim\left(CI_{W_i}(u_a, t_k), \ CI_{W_{i+1}}(u_a, t_k) \right) > k$

5.   Drifted V_1: = True

6.   if $(|sim\left(T_k, CI_{W_i}(u_a, t_k) \right) - sim\left(T_k, \ CI_{W_{i+1}}(u_a, t_k) \right)| > l$

             Drifted V_2: = True

7.   if (Drifted V_1 & Drifted V_2==True)

8. $U_{d_1} = U_{d_1} \cup \{u_a\}$

9.   Return U_{d_1}

3.2 Drift Detection Based on Kullback-Leibler Divergence

Layer 1 detection algorithm filters set of drifted users based on their information interest. We used layer 2 detection to further assess the users identified in layer 1. In layer 1, when apply the cosine similarity, we did not look at the individual corresponding features. Individual features may not close with each other even the cosine similarity is high. We need to check whether individual features are closed with each other. Kullback-Leibler (KL) Divergence [12] measures the relative entropy of two probability distributions. In this paper, we user KL divergence to measure the difference between a user's interest feature vectors over two time windows, which is calculated based on the individual corresponding features in the two interest feature vectors. In addition to comparing the user interest difference using Algorithm 2, we compared the probability distribution of user features over two consecutive time windows using KL Divergence which measures the relative difference of the distribution over two periods of time based on entropy. We compared the difference between a user's interest probability distribution over two windows, where size of the window is one month. Let $CI_{w_i}(u, t) = < p_{u,t,w}(v_1), p_{u,t,w}(v_2), \ldots, p_{u,t,w}(v_{|V|}) >$ be the feature distribution vector for a given user u for topic t in window w, $v \in V$, where V is a set of words in the entire posts. The difference between the user's interest probability distribution over two windows can be defined as follows w_i and w_{i+1} are two adjacent time windows. Equation 1 describes the KL divergence calculation for two adjacent time windows.

$$D(CI_{w_i}(u, t) | CI_{w_{i+1}}(u, t)) = \sum_{v \in V} p_{u,t,w_i}(v) log \frac{p_{u,t,w_i}(v)}{p_{u,t,w_{i+1}}(v)} \tag{1}$$

Algorithm 2 describes the steps of KL divergence calculation for two adjacent windows.

Algorithm 2: Layer 2 Drifted user detection based on KL Divergence

Input: Data for two time windows w_i, w_{i+1}
User set U_{d_1} returned from algorithm 1
User content profile CI,
Topic content representation T_j, $j = 1, \ldots n$,
Representative topics for each user $UT(u_i)$ $i = 1, \ldots, m$
experimental thresholds k

Output: Drifted User set U_{d_2}

---

1.   Set $U_{d_2} = \{\}$
2.   For each user u_a in U_{d_1}
3.   For each topic $t_k \in UT(u_a)$
    $kl := D(CI_{w_i}(u_a, t_k) | CI_{w_{i+1}}(u_a, t_k))$
         If (kl < k)
4. $U_{d_2} = U_{d_2} \cup \{u_a\}$
5.   Return U_{d_2}

The smaller divergence is an indication of no difference where a higher difference is an indication of a drift occurred.

3.3   Drift Detection Based on Feature Similarity Differences.

First two detection layers of our approach deal with user interest-based features. Existing spam detection approaches use both content-based features as well as non-content features. Therefore, in layer three detection we applied a set of features commonly used in the literature. Spammers will change their behavior to avoid detection. As a result, features used to detect spammers also get changed. We used the feature similarity difference in two time windows to identify drifted features. Then we apply KL Divergence to confirm the drifted features. Table 1 Summarizes features we used for the experiment. The feature set was extracted from the research [16].

Table 1. Features used for spam detection [16]

Feature name	Description
Number of Hashtags	Number of topics in posts denoted by #
Number of Mentions	Number of accounts mentioned in posts. Mentions are defined using @
Number of Retweets	Number of retweets for this post
Number of Spam words	Number of words which are labelled as spam words
Number of Trending Topics	Number of topics which are identified as tending topics by Twitter network
Number of hashtags per words	Ratio between number of topics and total number of words in posts
Number of URLs per Words	Ratio between number of URLs and total number of words in posts
Number of Numeric Characters	Ratio between number of digits present and total number of words in posts
Number of Replies	Number of replies received for a tweet
Tweet and Topic Content Divergence	Average KL Divergence between all posts of a given topic and content of the topic
Tweet and User's Tweets Divergence	Average KL Divergence between all posts of a user and other tweets
Number of Followee	Number of following users
Number of Follower	Number of follower users

First, we update the feature vectors based on their local feature differences in time windows. The method that we used is first introduced in [13]. We extended their approach with several modifications. As modifications, we applied a different set of features and we used KL divergence as an additional measure to detect the drift. Second, we compute the KL divergence difference between two feature vectors. First, we considered the similarity between two feature vectors of a user in two adjacent time windows. There is an experiment threshold used to determine the significance of the change. Second, for the set of filtered users, we applied KL Divergence for each selected user's feature vectors in two-time windows to detect the change occurred over time. Higher difference of KL divergence is not only an indication of a drift, but also an indication of a potential spam in OSN. Similarity difference and the KL divergence scores of each user would provide an indication of a potential drift but it is highly dependent on the nature of the features selected and the amount of new data available at the given time.

We developed an approach for drift detection by considering the changes of user's features in two consecutive time windows. The learning space consists of two parts, current and future and defined as S. Let $F = \{x_1, x_2, \ldots, x_n\}$ where x is the given feature and n is the total number of features in F. F is the feature vector of a user $u_i \in U$. Every user u_i is associated with a label, l where $l \in L$. $L = \{\text{Spam, Genuine}\}$. Therefore, the current learning space S_c can be defined as a vector, where $S_c =< F_{c_1}, l_{c_1}, \ldots, F_{c_n}, l_{c_n} >$. Similarly future learning space can be defined as $S_p = \{F_{p_1}, l_{p_1}, \ldots, F_{p_n}, l_{p_n}\}$. We calculated the cosine similarity between two learning spaces, current and future. If the similarity of the two feature vectors is above the given experimental threshold, we considered that, there is a drift of features in the future space. Therefore layer 3 will label a set of users who drifted based on non-user interest-based features. Final set of drifted users will be a union of two user sets output from layer 2 and layer 3 detections.

3.4 Validation of Drifted Users Based on Peer Acceptance

Validation is essential in hierarchical test-based drift detection. In this stage, we considered the peer acceptability of the user to verify the occurred drift. Research work, Peer Acceptance based Spam Detection (SDPA) [19] is used as a validation method to validate drifted users. Peer acceptability of the user should change when spammers drifted their interest/features. Peer Acceptance based spam detection system, detects the spammers based on their peer's acceptability in social network. We applied the peer acceptance-based spam detection algorithm to the latest window and received the potential spammers in that period. Then we cross checked them with the user sets identified in layer 2 and layer 3 detectors. If they have the same predicted label, we validate them as drifted users and update them with the new predicted label.

3.5 Drift Adaptation with Newly Predicted Labels

One of the main limitations in traditional classifier is the requirement of labelled datasets. When the users are drifted old labels used to train classifiers will not be significant for the new users. Therefore, it is essential to update the label of the training set. Our three-layered detection approach has identified the drifted users and they were validated through the validation layer. Finally, it predicts a new label for the drifted users through its learning process. In case of a significant difference in the features, the training set of the classifier is updated in a timely basis. As discussed earlier, the classifiers suffer with two problems of required labelled data and the updated labels for the changes. The update of the new knowledge is vital for the performance of classifiers. Therefore, when the concept drift is verified, we update the new feature values and retrain the classifier with the drifted features using the new predicted labels. We used Gradient Boost (GB), Naïve Bays (NB), Random Forest (RF) and Support Vector Machines (SVM) as the classification algorithms for validation and selected the best out of them. These classifiers are used to validate the drift detection in our experiment. First, we apply the classifier in old time window and obtain the results. Then we apply the classifier to new time window. It showed us, there are some users who identified as spammers in previous window have now identified as genuine users due to their drifted behaviors. Then we apply our drift learning algorithm to the new time window and retrain the classifiers with drifted features. Classifiers were applied again to test the performance. In the process of updating, we collect all user instances with their updated feature set and stored them in an intermediate repository. In case of a significant difference in the features, the training set of the classifier is updated in a timely basis with the new labels.

4 Experimental Setup

Our drift aware spam detection approach learns the drift from the user's content and detects spammers based on changes of features over time and users' peer acceptability using our proposed algorithms. The overall performance of the proposed approach achieved an accuracy difference of 0.21 with successfully addressing the drift occurred through different time windows.

4.1 Datasets and Criteria for Evaluation

We collected two twitter datasets for the period of 3 months.

- Peer Acceptance Data Collection (PADC) is a dataset collected in 2019 and consist of 4200 users. There are 2800 genuine users and 1400 spammers in total.
- User Information Interest dataset (UIID) is a dataset collected in 2019. It consists of 2400 users where 1500 of them are genuine and 900 spammers.

Both manual labelling and classification-based labelling is used to label the datasets.

Accuracy, precision, recall, and F1 score are used as criteria to evaluate the detection performance. Let TP denote true positive which is the number of correctly identified spammers, TN denote true negative which is the number of correctly identified genuine users, FP denotes false positive which is the number of genuine users who are wrongly identified as spammers, and FN denote false negative which is the number of spammers who are wrongly identified as genuine users, then, Accuracy = (TP + TN)/(TP + TN + FP + FN), Precision = TP/(TP + FP), Recall = TP/(TP + FN), F1 = 2*Precision*Recall/(Precision + Recall). The Error rate is also used to validate the correctness of updated features. There are different set of classifiers run for the dataset of first, time window and the Table 2 consists of the results for the classifiers with drifting, without the effect of drifting and after considering the drift.

4.2 Experiment Results and Evaluation

There are four different classifiers, Random Forest, SVM, Gradient Boost and the Naive Bayes classifiers used for the evaluation. The results clearly show that our model can perform better than the existing classifiers when drifted. The classifiers used for the old-time window (w_i) will not be able to perform similarly for the current time window (w_{i+1}) and the performance were decreased. The drift detection model has successfully handled the drifting and the performance difference from the previous time window to current is 0.21 in terms of best performance. Table 2 consists of the best results achieved for two-time windows.

Table 2. Performance comparison of classifiers and our model with drifting

Dataset	Classifier	Classifier result (w_i)	Classifier result with drift (w_{i+1})	Our result (w_{i+1})	Difference from previous window
PADC	**GB**	98.82	**96.33**	**98.61**	**0.21**
	SVM	98.12	96.07	97.66	0.46
	RF	98.31	96.69	97.88	0.43
	NB	97.64	96.21	97.04	0.60
UII	**GB**	98.53	**96.53**	**98.22**	**0.31**
	SVM	98.16	95.98	97.71	0.45
	RF	97.89	96.17	97.43	0.46
	NB	97.45	96.06	96.91	0.54

5 Discussion

As described earlier, there are two main types of concept drifts, real drift, and virtual drift. The similarity-based measurement is an ideal solution to detect the virtual drift in OSN users but detection of real drift using similarity-based approach would be time consuming and more computationally intensive. Researchers in concept drifting have introduced different approaches for such real concept drift detection [13, 16, 17].

Therefore, to deal with the real concept drift in a more powerful and efficient manner, we combined the similarity-based measurements with KL divergence to detect the drift occur in user's space. Similar to FSM, we compared the difference between two spaces, current and future. We used the same feature set in FSM and calculated the KL divergence scores for the two feature distributions. We use two different types of features, user's information interest-based features and a set of commonly used features in the literature. Finally for every user instance, we have two values computed from FSM and KLDM. For the instances, which were temporally, selected as drifted we applied the SDPA, and validate with the detection label in SDPA. The set of features that we used for spam drift detection in addition to the features described in Table 1 are as follows. F1: Calculate same user's content similarity of each topic across multiple time windows. If the similarity value is quite high or low, that can be a spammer. For general users, their similarity values do not have such a behavior. F2: we assumed that the topic content $\overrightarrow{T}_t$ similarity between two windows must be consistent with the user's content interest in two-time windows. A considerable deviation is an indication of drifting of the content. Then we Calculate the user's content interest for same two-time windows for the same topic t. A large deviation is an indication of a suspicious behavior. A continuous variation across multiple windows is a clear indication of spamming behavior. F3: It is noticed from the analysis that, usually the percentage of frequent topics used by a certain user is remains between 10–25% for general users. Their frequent topic usage is varied between time windows. Spammers have used more frequent topics. In contrast, for a given window, spammers frequent topic usage is between 25%–45%. The spammers frequent topic usage is usually consistent in time windows. If a user, maintain a higher frequent topic usage across multiple windows and if their frequent topic usage percentage is consistent, it is a suspicious behavior. F4: Frequent usage of a topic by a certain user over multiple windows in multiple messages. If a hash tag is continuously used by a certain user, it is suspicious. There are no additional computations involved in these four features, since they are calculated with the generation of peer acceptance. Once all three levels of drift detection are passed, the selected instances were passed to the transfer learning module to replace existing instances with the newly drifted instances.

Finally, the classifier will be retrained with newly learnt features and new labels for the spam detection. The feature selection for the drift aware spam detection was challenging where we used a mixture of our own features with features found in the literature. The investigation reveals that some of the features used in literature review are not justified in terms of real time suitability where we need to analyze the real time suitability of the features. It gave us the opportunity to select some features, which are directly extracted from users post content, without complex processing, yet a powerful set of features for spam detection. The threshold selection for both FSM and KLDM was challenging since the feature space is quite large across windows. We have considered the average similarity of feature group across all windows as the threshold to determine the drift occurred across the feature spaces. More experiments could have conducted to fine tune the threshold to reduce the error rate and improve the classifier performances.

Selection of classifiers for the evaluation also an important task. We did not use any data streaming classifiers since our aim is to satisfy the requirement of relabeling and training the non-streaming classifiers with the drift. The evaluation of the drift treatment also challenging since there is no labelled data available for public datasets for the latest versions. Use of old labels for newer datasets might be little unfair for the systems, but we tried to use new data to retrain the classifiers to learn the new label.

6 Conclusion

The paper presents a novel approach for spam drift detection in online social networks. First it investigated the nature of a set of spam features used in the literature and analyzed the real time suitability of these features for the spam drift detection. It presents a hybrid approach of combating social spammers where it handles the spam drift in an unsupervised manner. The concept drift occurred in OSNs are detected using set of features learn and the feature similarity, User's Information interest, KL divergence and the Peer acceptability is used to automatically update the drift occurred in the dataset. The new knowledge will be then used to update the existing classifier model and retrain the classifier for new detection. The results are encouraging, and our model is capable of handling the spam drift better than the retrain of classifiers with the existing labels.

References

1. Widmer, G., Kubat, M.: Learning in the presence of concept drift and hidden contexts. Mach. Learn. **23**, 69–101 (1996)
2. Krawczyk, B., Minku, L., Gama, J., Stefanowski, J., Woźniak, M.: Ensemble learning for data stream analysis: a survey. Inf. Fusion. **37**, 132–156 (2017)

3. Huang, G., Zhu, Q., Siew, C.: Extreme learning machine: theory and applications. Neuro-computing **70**, 489–501 (2006)
4. Yu, S., Wang, X., C. Príncipe, J.: Request-and-reverify: hierarchical hypothesis testing for concept drift detection with expensive labels. In: Proceedings of the Twenty-Seventh International Joint Conference on Artificial Intelligence (2018)
5. Lu, J., Liu, A., Dong, F., Gu, F., Gama, J., Zhang, G.: Learning under concept drift: a review. IEEE Trans. Knowl. Data Eng. **31**(12), 1–1 (2019). https://doi.org/10.1109/TKDE.2018.287 6857
6. Wu, T., Wen, S., Xiang, Y., Zhou, W.: Twitter spam detection: survey of new approaches and comparative study. Comput. Secur. **76**, 265–284 (2018)
7. Rao, S., Verma, A., Bhatia, T.: A review on social spam detection: Challenges, open issues, and future directions. Expert Syst. Appl. **186**, 115742 (2021)
8. Gama, J., Žliobaitė, I., Bifet, A., Pechenizkiy, M., Bouchachia, A.: A survey on concept drift adaptation. ACM Comput. Surv. **46**, 1–37 (2014)
9. Sethi, T., Kantardzic, M.: On the reliable detection of concept drift from streaming unlabeled data. Expert Syst. Appl. **82**, 77–99 (2017)
10. Sethi, T., Kantardzic, M., Arabmakki, E.: Monitoring classification blindspots to detect drifts from unlabeled data. In: 2016 IEEE 17th International Conference on Information Reuse and Integration (IRI) (2016)
11. Biggio, B., et al.: Evasion attacks against machine learning at test time. In: Blockeel, H., Kersting, K., Nijssen, S., Železný, F. (eds.) ECML PKDD 2013. LNCS (LNAI), vol. 8190, pp. 387–402. Springer, Heidelberg (2013). https://doi.org/10.1007/978-3-642-40994-3_25
12. Kullback, S., Leibler, R.: On information and sufficiency. Ann. Math. Stat. **22**, 79–86 (1951)
13. Henke, M., Santos, E., Souto, E., Santin, A.: Spam detection based on feature evolution to deal with concept drift. JUCS J. Univ. Comput. Sci. **27**, 364–386 (2021)
14. Cao, C., Caverlee, J.: Behavioral detection of spam URL sharing: Posting patterns versus click patterns. In: 2014 IEEE/ACM International Conference on Advances in Social Networks Analysis and Mining (ASONAM 2014) (2014)
15. Lu, N., Lu, J., Zhang, G., Lopez de Mantaras, R.: A concept drift-tolerant case-base editing technique. Artif. Intell. **230**, 108–133 (2016)
16. Washha, M., Qaroush, A., Mezghani, M., Sedes, F.: Unsupervised collective-based framework for dynamic retraining of supervised real-time spam tweets detection model. Expert Syst. Appl. **135**, 129–152 (2019)
17. Gama, J., Medas, P., Castillo, G., Rodrigues, P.: Learning with drift detection. In: Bazzan, A.L.C., Labidi, S. (eds.) SBIA 2004. LNCS (LNAI), vol. 3171, pp. 286–295. Springer, Heidelberg (2004). https://doi.org/10.1007/978-3-540-28645-5_29
18. Gama, J., Castillo, G.: Learning with local drift detection. In: Li, X., Zaïane, O.R., Li, Z. (eds.) ADMA 2006. LNCS (LNAI), vol. 4093, pp. 42–55. Springer, Heidelberg (2006). https://doi.org/10.1007/11811305_4
19. Niranjan Koggalahewa, D., Xu, Y., Foo, E.: Spam detection in social networks based on peer acceptance. In: Proceedings of the Australasian Computer Science Week Multiconference (2020)
20. Frias-Blanco, I., Campo-Avila, J., Ramos-Jimenez, G., Morales-Bueno, R., Ortiz-Diaz, A., Caballero-Mota, Y.: Online and non-parametric drift detection methods based on Hoeffding's bounds. IEEE Trans. Knowl. Data Eng. **27**, 810–823 (2015)
21. Lu, N., Zhang, G., Lu, J.: Concept drift detection via competence models. Artif. Intell. **209**, 11–28 (2014)
22. Shao, J., Ahmadi, Z., Kramer, S.: Prototype-based learning on concept-drifting data streams. In: Proceedings of the 20th ACM SIGKDD International Conference on Knowledge Discovery and Data Mining (2014)

23. Alippi, C., Boracchi, G., Roveri, M.: Hierarchical Change-Detection Tests. IEEE Trans. Neural Netw. Learn. Syst. **28**, 246–258 (2017)
24. Xu, N., Huo, C., Zhang, X., Cao, Y., Meng, G., Pan, C.: Dynamic camera configuration learning for high-confidence active object detection. Neurocomputing **466**, 113–127 (2021)
25. Taoufik, N., Boumya, W., Achak, M., Chennouk, H., Dewil, R., Barka, N.: The state of art on the prediction of efficiency and modeling of the processes of pollutants removal based on machine learning. Sci. Total Environ. **807**, 150554 (2021)
26. Zheng, J., Cole, T., Zhang, Y., Kim, J., Tang, S.: Exploiting machine learning for bestowing intelligence to microfluidics. Biosens. Bioelectron. **194**, 113666 (2021)
27. Tsai, C., Chen, Y., Tang, T., Luo, Y.: An efficient parallel machine learning-based blockchain framework. ICT Express **7**, 300–307 (2021)

Hospital Readmission Prediction Using Semantic Relations Between Medical Codes

Sea Jung Im$^{(\boxtimes)}$ (iD), Yue Xu$^{(\boxtimes)}$ (iD), and Jason Watson$^{(\boxtimes)}$ (iD)

Queensland University of Technology Brisbane, Brisbane, Australia
seajung.im@hdr.qut.edu.au, {yue.xu,ja.watson}@qut.edu.au

Abstract. Unexpected hospital readmissions are problematic to both hospitals and patients. Prediction of patients' readmission becomes an important task. Recurrent neural networks (RNN) and the attention mechanisms have been proposed to learn temporal relationships between patient' admissions for readmission prediction. Existing works demonstrate that incorporating medical ontologies can be beneficial to prediction tasks. However, it ignores the importance of semantic information of medical codes which can be found in the codes' descriptions. Therefore, we propose a model called Code Description Attention Model (CDAM), which adopts codes' descriptions into readmission prediction model via RNN and the attention mechanisms to explore the semantic information about medical codes. Experimental results show that CDAM improves not only the performance of readmission prediction but also the quality of codes' embeddings.

Keywords: Hospital readmission · Medical ontology · Code description · Semantic relationship

1 Introduction

A hospital readmission is considered when a patient returns to the same hospital or other hospitals within a certain time frame after discharge from the initial admission. Unexpected readmissions cost an extra huge amount of money on hospitals, which raises the great demand for reducing them [15]. One of the effective way of preventing hospital readmission is to predict patients with high risks [14]. The increased adoption of Electronic Health Records (EHR) which contains an enormous amount of information about patient admission attracts growing interests in applying deep learning techniques in healthcare data as they have proven to be effective to handle complex data like EHR .

Recurrent neural network (RNN) has been commonly applied to analyse EHR because it allows capture long dependencies between sequences and EHR contains long sequences of patients' visits [1,2,5,9,11]. Choi et al. [1] use RNN on reversed sequential data and Ma et al. [9] employed bidirectional RNN for diagnosis prediction in next visit.

© Springer Nature Singapore Pte Ltd. 2021
Y. Xu et al. (Eds.): AusDM 2021, CCIS 1504, pp. 62–71, 2021.
https://doi.org/10.1007/978-981-16-8531-6_5

Skip-gram proposed by Mikolov et al. [10] is commonly applied to capture co-occurrence information in EHR [3,6]. However, co-occurrences may not be able to capture intrinsic relationship between medical concepts. Recently, Choi et al. [4] proposed the GRAM model to exploit the hierarchical structures of medical ontologies such as International Classification of Diseases (ICD) [8] or Clinical Classifications Software(CCS). However, it only considers the structure of relationships between diagnosis codes and ignore the importance of semantic meaning. We believe that semantic meanings of medical codes are important to determine the relationships between codes.

For example, ICD-9 code 585.1, 585.2 and 585.6 share the same parent "Chronic kidney disease". The definition of two codes 585.1 and 585.2 are "Chronic kidney disease stage 1" and "Chronic kidney disease stage2", which shows similarity. However, the definition of the code 585.6 is "End stage renal disease", which is completely different from the two codes. GRAM [4] would not tell the difference between them because it only considers the structure. Specifically, the description associated with the medical codes in the medical ontology provide the source to find the semantic information and the multi-occurrence of words in the descriptions is particularly useful capture the meaning of medical codes. Therefore, we propose to explore the semantic information of medical codes from a medical ontology and take the semantic information into consideration to represent medical codes and the relationships between them more accurately.

The main contributions of this study are summarised as follows: 1) to propose the usage of definitions of medical codes in ontology to learn semantic meaning of diagnosis codes 2) to consider the repetition of words in descriptions to reflect the significance of repeated words to generate improved representation for medical codes.

2   Related Works

Recently, many works on medical predictive tasks use embedding techniques [10]. The key point behind word embedding is that medical codes frequently occurring together will show that these codes have similar meanings. Choi et al. [2] and Ma et al. [9] employ Skip-gram to capture the co-occurrence relationships between medical codes in EHR. However, this approach only considers co-occurrence and does not take sequential order of medical codes. Choi et al. [3] propose Med2Vec to learn both sequential order information and co-occurrence of the codes.

RNN and attention mechanism have been used to model temporal relationships between medical events. Dipole [9] employed bi-directional RNN on three different attention mechanisms and RETAIN [1] implements reverse time attention mechanisms. GRAM [4] also utilises both RNN and attention mechanism. Xiao et al. [14] propose a hybrid TopicRNN which combines topic modeling and RNN.

Recently, the GRAM model proposed in [4] leverages hierarchical structure to improve the quality of representations of medical codes. By employing parent-child relationship, it generates representations that reflect domain knowledge from ontology. This means that diagnosis codes with the same ancestors have similar representations. The attention mechanism decides which ancestor of the code is given the most weight. However, GRAM only uses the structure of medical codes that act as independent codes and does not consider any semantic meaning of the codes. In fact, the medical ontologies not only provide the relationships of medical concepts but also they provide semantic information about medical codes or concepts via descriptions to those medical codes or concepts. We believe that incorporating the definitions of medical codes into prediction model can improve the quality of prediction performance as well as interpretation of the model results.

3   Proposed Model

In this section, we describe the details of our proposed model. We aim to generate more accurate representations of diagnosis codes by incorporating the description of diagnosis codes in a given medical ontology such as ICD or CCS.

3.1   Basic Notation

In this paper, we denote the entire set of medical codes from EHR including diagnosis and procedure codes as $c_1, c_2, c_3 .... c_{|C|} \in C$ with the total number of medical codes $|C|$. Each patient can have multiple admissions $A_1, A_2 ...., A_T$ where each admission consists of a subset of medical codes $A_t \subseteq C$. Each admission A_t can be represented as a binary vector $x_t \in \{0, 1\}^{|C|}$ where the i_{th} element of x_t is 1 if the code c_i occurs in the admission A_t, otherwise, 0.

The given medical ontology can be represented as a graph where all nodes form a set $N = C \bigcup C\prime \bigcup W$. C consists of all diagnosis and procedure codes which are the leaf nodes. The set $C\prime = \{c_{|c|+1}, c_{|c|+2}, ..., c_{|c|+|c\prime|}\}$ consists of all non-leaf nodes (i.e. ancestors of the leaf nodes). The set $W = \{w_1, w_2, w_3, ..., w_{|w|}\}$ consists of all unique words from the associated descriptions.

3.2   Generate Description Features

The ontology ICD is represented as a graph which shows the parent-child relationship between nodes. The model GRAM [4] utilises this hierarchical structure of the ontology to learn representations of diagnosis and procedure codes. However, GRAM does not consider the semantic meaning of the codes and in this research, we propose to take medical codes' descriptions, that contains more semantic meanings than codes themselves, into consideration in learning the code representations.

The description of a node consists of one or more sentences. Not all the words in a description are useful to represent the meaning of the node, e.g.,the stopping words such as "the", "a",etc. To keep the descriptions more meaningful, we first remove all stop words and then apply pattern mining methods to select more representative words, which are called "description features", to represent descriptions.

The description features for each node are generated based on the node's descriptions of its children nodes. For p ∈ N, we construct a local transactional dataset which consists of the descriptions of p's children codes and the description of p. Before building the transactional dataset, we remove stopping and highly common words. Let $w(p)$ be the set of words in p's description and $W(p) = w(p) \bigcup_{c \in children(p)} w(p)$, where children(p) is a set of p's children. $W(p)$ is the set of candidates for finding the features to represent p's description. By applying pattern mining methods, we can generate the relative support for each word in W(p). If the relative support is greater than a given threshold, the word is selected as a description feature of p. DF(P) denotes all the selected description features of p.

$$DF(p) = \{w | sup(w) > \theta, w \in W(p)\} \tag{1}$$

where θ, $0 \leq \theta \leq 1$ is a threshold.

Take an ICD-9 code 580–629 as an example. Table 1 shows an example of hierarchical structure of ICD-9 code 580–629, which has 6 children nodes, 580–589, 590–599, 600–608, 610–612, 614–616, and 617–629. w(580–629) = {Genitourinary}, w(580–589) = {Nephritis, Nephrotic, Nephrosis}, etc.. W(580–629) = {Genitourinary, Nephritis, Nephrotic, Nephrosis, Urinary, Male, Genital, Organs, Breast, Heart, Inflammatory, Female, Pelvic, Organs, Female, Genital, Tract}

Let θ set to be 0.3. When generating description features, words from a associated code are include as description features. Therefore, "Genitourinary" is a description feature. The words "Genital", "Organs" and "Female" appear two times, which means that its relative support are about 0.33 (2 out of 6). Therefore, these words are selected as description features for the code (580–629).

Table 1. An example of hierarchical structure of the ICD-9 code

Level	ICD-9	Description
0	Root	Root
1	580–629	Diseases of the genitourinary system
2	580–589	Nephritis, nephrotic syndrome, and nephrosis
2	590–599	Other diseases of urinary system
2	600–608	Diseases of male genital organs
2	610–612	Disorders of breast
2	614–616	Inflammatory disease of female pelvic organs
2	617–629	Other disorders of female genital tract

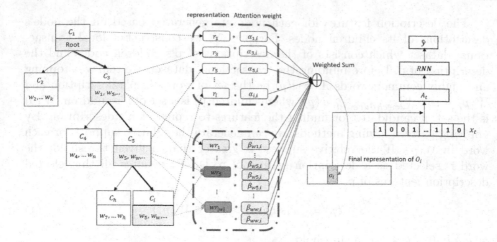

Fig. 1. The illustration of the proposed model. Medical codes in the ontology are denoted $c_1, c_2, c_3 .... c_{|C|}$ and description features are denoted as $w_1, w_2, w_3, ..., w_{|w|}$. Final representation o_i is generated by combining the basic representations of code and description features with attention mechanism. The final embedding matrix is multiplied with an admission x_t to generate a visit representation A_t which is fed to the prediction model.

3.3   Learning Representations with description Features

We propose an attention model, called Code Description based Attention Model (CDAM), for learning medical code representation based on code descriptions. Figure 1 represents the overall of our proposed model. The left part of the figure describes the structure of the medical ontology ICD. The upper part in the centre shows the attention mechanism with ancestors of a code c_i and the low part is the attention mechanism with description features. RNN prediction model is displayed in the right part of Fig. 1. In Fig. 1, a node represents a medical code and each node consists of a code and description features generated by the previous step. Each node c_i has a learnable embedding $r_i \in \mathbb{R}^m$, where m represents the dimension size of the embedding, the code embedding is also called the code representation. The final embedding of a medical code o_i can be obtained by combining of the basic representations of itself, its ancestors and description features' representations as given below.

$$o_i = \sum_{j \in \mathcal{A}(i)} \alpha_{ij} r_j + \sum_{j \in \mathcal{A}(i)} \sum_{w_f \in \mathcal{DP}(c_j)} \beta_{if} r_f \tag{2}$$

where

$$\left(\sum_{j \in \mathcal{A}(i)} \alpha_{ij} + \sum_{j \in \mathcal{A}(i)} \sum_{w_f \in \mathcal{DP}(c_j)} \beta_{if} \right) = 1 \tag{3}$$

where o_i represents the final representation of diagnosis code c_i. j denotes the index of ancestors of c_i and $A(i)$ contains the indices of c_j's ancestors and c_i

itself. α_{ij} denotes the attention weights of c_j's acestor c_j for c_i and β_{if} denotes attention wights for description features w_f for a code c_j. f denotes the index of description features of c_i. As seen in Fig. 1, some description features may occur in multiple nodes within the same path. For example, w_5 appears in c_3, c_5, c_i and our proposed model will include the representation of w_5 three times when updating the representation of o_i. As a result, our model can emphasis the importance of repeated description features.

The attention weights α_{ij} and β_{if} are calculated by the following equations

$$\alpha_{ij} = \frac{\exp(\theta(r_i, r_j))}{\sum_{k \in \mathcal{A}(i)} \exp(\theta(r_i, r_k))} \tag{4}$$

,

$$\beta_{if} = \frac{\exp(\theta(r_i, r_f))}{\sum_{k \in \mathcal{A}(i)} \sum_{g \in \mathcal{DP}(c_k)} \exp(\theta(r_i, r_g))} \tag{5}$$

$\theta(r_i, r_k)$ is a scalar value and defined as

$$\theta(r_i, r_j) = u_a^T \tanh \left(W_a \begin{bmatrix} r_i \\ r_j \end{bmatrix} + b_a \right), \tag{6}$$

where the weight matrix $W_a \in \mathbb{R}^{d*2m}$, the bias vector $b_a \in \mathbb{R}^d$ and the weight vector $u_a \in \mathbb{R}^d$ and are the parameters to be learned. The d represents the dimension size of $\theta(r_i, r_k)$.

3.4 Prediction Model

Recurrent neural networks (RNN) is adopted as a predictive model for readmission. We concatenate final representations of all medical codes and then generate the embedding matrix $O \in \mathbb{R}^{m*|c|}$. With an hospital admission A_t is represented as a binary vector x_t, we multiply embedding matrix O and x_t to convert A_t to a visit representation a_t. Then, the visit representation v_t is fed into RNN predictive model.

4 Experiments

We conduct experiments to evaluate the predictive performance of our CDAM in terms of readmission and the quality of code representations by comparing that of GRAM [4].

4.1 Experimental Setup

Datasets. We conduct experiments on two real-world datasets: MIMIC-III (Medical Information Mart for Intensive Care) dataset and CKD dataset. Table 2 shows the basic statistics of two datasets.

MIMIC-III Dataset. is an open-source dataset from the Beth Israel Deaconess Medical Centre in Boston, Massachusetts containing extensive data about patients over 11 years [7].

CKD Dataset. contains hospital admission data for a group of patients with chronic kidney diseases (CKD) receiving renal specialty care in the public health system in Queensland, Australia between 2011 and 2018 [13].

Table 2. Basic statistics of MIMIC-III and CKD datasets

Dataset	MIMIC-III	CKD
& of patients	7,499	6,772
& of visits	19,911	77,342
Avg. & of visits per patient	2.66	11.42
& of unique ICD codes	6,419	7,990
Avg. & of codes per visit	17.11	8.51
Max & of codes per visit	69	142

Data Preprocessing. We extract records of patients with at least two admissions where diagnosis codes exist. When an admission occurs within 30 days from the last discharge, it is labeled as a readmission, 1, otherwise, 0.

Implementation Settings. Diseases are represented with ICD-9 codes in the MIMIC-III dataset and with ICD-10 in the CKD dataset. Therefore, we use the hierarchical structure of ICD-9 and ICD-10 ontologies for MIMIC-III and CKD respectively. We use Area under the ROC curve (AUC) to measure the prediction performance. We conducted experiments with 10-fold cross-validation by splitting datasets into training set, validation set and test set, and fix the size of validation set and test set to be 10%. We vary the size of the original dataset from 20% to 80% to evaluate the model's performance with insufficient training data. All models are optimized using Adam with an initial learning rate of 0.001 and the batch size is fixed to 100. The coefficient of L2 norm regularization is fixed to 0.001. Hyperparameters are defined as in MIMIC-III, dimension of embedding size, dimension of RNN hidden layer, and dimension of W_a, b_a l from the Eq. (6) are 200, 200 and 100 respectively and in CKD, they are 100, 100, and 100. Dropout rate for dropout on the RNN hidden layer is defined 0.4 for MIMIC-III and 0.2 for CKD.

4.2   Experimental Results

Readmission Prediction. We first evaluate the performance of readmission prediction. Table 3 shows the readmission prediction performance on the two data sets, MIMIC-III and CKD. On MIMIC-III dataset, CDAM outperforms the baseline model GRAM regardless of the size of training set and showed maximum 6.6% of improvement AUC. On CKD dataset, the performance of CDAM on 80% and 60% of the training set shows minor improvement but it is slightly worse than GRAM on 40% and 20% of the training set.

Table 3. AUC of readmission prediction on MIMIC-III and CKD. We conduct experiments by varying the size of training data from 20% to 80%.

Dataset	Model	80%	60%	40%	20%
MIMIC-III	GRAM	0.6776	0.6414	0.6725	0.6585
	CDAM	**0.6835**	**0.6839**	**0.6853**	**0.6625**
CKD	GRAM	0.6996	0.6893	**0.6914**	**0.6912**
	CDAM	**0.7024**	**0.6949**	0.6864	0.6846

Representation of Medical Codes. We also evaluate how well the medical codes are represented by the generated code embeddings. Based on the generated embeddings, the medical codes are clustered into groups. With good embeddings, similar codes would be clustered into the same group. We evaluate the quality of the code embeddings by evaluating the quality of the clusters using Silhouette score [12] which is calculated by measuring the mean distance within a cluster and the mean distance between clusters. The value lies between -1 and +1, which +1 indicates the best value. Table 4 shows the result of silhouette score on both MIMIC-III and CKD based on the number of clusters. The comparison shows that our representation outperforms that of GRAM.

Table 4. Silhouette score on MIMIC-III and CKD.

Dataset	Model	Number of clusters		
		10	20	30
MIMIC-III	GRAM	0.049	0.057	0.072
	CDAM	**0.063**	**0.089**	**0.105**
CKD	GRAM	0.053	0.078	0.083
	CDAM	**0.081**	**0.093**	**0.115**

5 Conclusion

In this research, we propose the adoption of descriptions of medical codes in a given medical ontology in addition to the parent-child relationship. Most of current works on healthcare prediction tasks only focus on the structure of medical ontologies or co-occurrence in the dataset, which ignores the importance of semantic meaning in the description of codes. Our proposed method can not only reflect the semantic meaning of medical codes in the ontology but also the significance of repeated words in the same path. Our method shows improved predictive performance of readmission and generates better representations of medical codes.

References

1. Choi, E., Bahadori, M.T., Kulas, J.A., Schuetz, A., Stewart, W.F., Sun, J.: Retain: an interpretable predictive model for healthcare using reverse time attention mechanism. In: Advances in Neural Information Processing Systems, pp. 3504–3512 (2016)
2. Choi, E., Bahadori, M.T., Schuetz, A., Stewart, W.F., Sun, J.: Doctor AI: predicting clinical events via recurrent neural networks (2015)
3. Choi, E., et al.: Multi-layer representation learning for medical concepts. In: Proceedings of the 22nd ACM SIGKDD International Conference on knowledge discovery and data mining. KDD '16, vol. 13–17, pp. 1495–1504. ACM (2016)
4. Choi, E., Bahadori, M.T., Song, L., Stewart, W., Sun, J.: Gram: Graph-based attention model for healthcare representation learning. In: Proceedings of the 23rd ACM SIGKDD International Conference on Knowledge Discovery and Data Mining, KDD '17, pp. 787–795. ACM (2017)
5. Esteban, C., Staeck, O., Baier, S., Yang, Y., Tresp, V.: Predicting clinical events by combining static and dynamic information using recurrent neural networks. In: Proceedings - 2016 IEEE International Conference on Healthcare Informatics, ICHI 2016, pp. 93–101 (2016)
6. Feng, Y., et al.: Patient outcome prediction via convolutional neural networks based on multi-granularity medical concept embedding. In: 2017 IEEE International Conference on Bioinformatics and Biomedicine (BIBM), pp. 770–777. IEEE (2017)
7. Goldberger, A.L., et al.: Physiobank, physiotoolkit, and physionet components of a new research resource for complex physiologic signals. Circulation 101(23), e215–e220 (2000). https://doi.org/10.1161/01.CIR.101.23.e215
8. of Library, W.H.O.O., Services, H.L.: Styles for bibliographic citations : guidelines for who-produced bibliographies (1988)
9. Ma, F., Chitta, R., Zhou, J., You, Q., Sun, T., Gao, J.: Dipole: diagnosis prediction in healthcare via attention-based bidirectional recurrent neural networks. In: Proceedings of the 23rd ACM SIGKDD International Conference on Knowledge Discovery Data Mining, KDD '17, vol. 129685, pp. 1903–1911. ACM (2017)
10. Mikolov, T., Chen, K., Corrado, G., Dean, J.: Efficient estimation of word representations in vector space (2013). arXiv.org
11. Pham, T., Tran, T., Phung, D., Venkatesh, S.: Deepcare: a deep dynamic memory model for predictive medicine (2016)
12. Rousseeuw, P.J.: Silhouettes: a graphical aid to the interpretation and validation of cluster analysis. J. Comp. Appl. Math. 20, 53–65 (1987). https://doi.org/10.1016/0377-0427(87)90125-7
13. Venuthurupalli, S.K., Hoy, W.E., Healy, H.G., Cameron, A., Fassett, R.G.: CKD.QLD: establishment of a chronic kidney disease [CKD] registry in Queensland, Australia. BMC Nephrol. 18(1), 189 (2017). https://doi.org/10.1186/s12882-017-0607-5, https://www.ncbi.nlm.nih.gov/pubmed/28592254

14. Xiao, C., Ma, T., Dieng, A.B., Blei, D.M., Wang, F.: Readmission prediction via deep contextual embedding of clinical concepts. PLOS One **13**(4), e0195024 (2018). https://doi.org/10.1371/journal.pone.0195024, https://www.ncbi.nlm.nih.gov/pubmed/29630604
15. Zheng, B., Zhang, J., Yoon, S.W., Lam, S.S., Khasawneh, M., Poranki, S.: Predictive modeling of hospital readmissions using metaheuristics and data mining. Expert Syst. Appl. **42**(20), 7110–7120 (2015). https://doi.org/10.1016/j.eswa.2015.04.066, http://www.sciencedirect.com/science/article/pii/S0957417415003085

HFM++: An Enhanced Holographic Factorization Machine for Recommendation

Zhengxin Fang[1], Mingcheng Qu[1], Shuai Zhang[2], Jingyu Zhang[5], Yi Yuan[4], Lina Yao[3], and Shiping Chen[5(✉)]

[1] Faculty of Computing, Harbin Institute of Technology, Harbin, China
19s103227@stu.hit.edu.cn, qumingcheng@hit.edu.cn
[2] ETH Zurich, Zürich, Switzerland
[3] School of Computer Science and Engineering, University of New South Wales, Kensington, Australia
lina.yao@unsw.edu.au
[4] School of Engineering, University of Sydney, Sydney, Australia
yyua8222@uni.sydney.edu.au
[5] CSIRO Data61, Sydney, Australia
{Jingyu.Zhang,Shiping.Chen}@data61.csiro.au

Abstract. Recommender systems are essential and are playing a more and more important role in our daily life, ranging from entertainment to online shopping. They have great commercial value, not only can improve the user experience by saving users time to locate related items, but also increase the exposure rate of long-tail items. Factorization machines (FMs) have been serving as an effective go-to recommender algorithm for years. Yet, it is not without disadvantages, such as mediocre performances. In this paper, we propose an enhanced factorization model, namely, enhanced holographic factorization machines (or HFM++). Our model integrates a Restricted Boltzmann Machine (RBM) and a generalized matrix factorization (GMF) into holograph FM, resulting in a more powerful recommendation model that inherits the benefits of all these three models. To evaluate the effectiveness and accuracy of HFM++, we conduct experiments with several well-known public benchmarking datasets and compare them against a number of baseline models. The experimental results show that HFM++ obtains the best performances across all metrics, outperforming recently published baselines.

Keywords: Neural matrix factorization · Factorization machine · Holographic factorization machines · Restricted Boltzmann Machine · Collaborative filtering

1 Introduction

With the rise of web services such as YouTube, Netflix, Amazon, consumers are suffering from over-choice. Identifying the most suitable items from millions of items is a miserable and even impossible task. Fortunately, recommender systems make it much easier by generating recommendations in alignment with users' preferences. It is fair to say that recommender systems [12,19,20] have become indispensable in our daily life [11,22]. We rely heavily on it while shopping online, looking for movies, songs, places-of-interest, and even finding friends, connections on social media. In the meanwhile, companies are boosting their profit by deploying recommendation algorithms as it can stimulate transactions and retain customers, helping them stand out from competitors. The advertising industry also uses it for advertisement recommendations based on users' personalized interests. As such, it is important to enhance the performance of recommender systems [2,27,28].

Across the rich history of recommender systems research, techniques based on matrix factorization (MF) [13] were highly dominant. The key point of the matrix factorization model is to factorize a user-item interaction matrix and learn the latent patterns of user behaviors and item characteristics. The limitation (e.g., lack of expressiveness and mediocre performances [7]) of matrix factorization is obvious. So, many research efforts were put to enhance matrix factorization. One of the most popular algorithms is factorization machines (FMs) [16]. Factorization machine is a generalization of matrix factorization and it provides the capability in modeling the pairwise interactions between features automatically and efficiently. FMs have demonstrated widespread success both in several recommendation tasks. However, FMs adopts inner product as the tool for interaction modeling, which might not be able to capture the complex structure of user interaction data as it is essentially a linear operator. Moreover, the summation operation adopted by FMs may result in future information loss.

To remit the aforementioned weakness, this paper presents an enhanced FM architecture, namely HFM++. HFM++ integrates generalized matrix factorization (GMF) [7], holographic factorization machines (HFM) [25], and Restricted Boltzmann Machine (RBM) [21] seamlessly in an end-to-end paradigm. GMF is a component of the neural collaborative filtering (NCF) framework [7] and it generalizes and extends the MF model into a nonlinear space with nonlinear activations. HFM is an improved FM model with better representation capability. In HFM, the inner product is replaced with holographic reduced representations (HRRs) [15]. More specifically, each input feature vector is modeled with a memory trace and individual feature interactions can be efficiently retrieved by an associative retrieval mechanism. Our model takes the best of both worlds by combining the outputs of GMF and HFM in the first place. Afterward, the result is fed into an RBM model. We adopt RBM because of its strong representation capability and non-linearity. Overall, we hypothesize that a combination of GFM, HFM, and RBM can lead to higher expressiveness and performance improvement as it inherits the benefits of all these three models, and our experiment also confirmed this hypothesis.

To summarize, the main contributions of this work are as follows:

- We propose an enhanced holographic factorization machine - HFM++, which is advantageous in modeling linearity, non-linearity, and complex feature interactions.
- We conduct experiments on four benchmark datasets and demonstrate that HFM++ obtains a relative performance improvement of 1% - 12% across four important metrics (rou_auc_score, accuracy, precision, and recall), outperforming recent models such as AFM [26] and HFM.
- We also conduct model analysis to inspect the impact of certain hyperparameters choices and model components.

This paper was organized as follows: In Sect. 2, we present the relevant background knowledge and technology notations used in the paper. In Sect. 3, our model is presented in detail. Section 4 is the experiment and the experiment result, and Sect. 5 is the relative work of this topic. At last, Sect. 6 is the conclusion of this paper.

2 Preliminary

In this section, we present the relevant background knowledge and notations used in the paper.

Table 1. Notation and descriptions

Notation	Meaning
v_i	The i-th dimension of the latent vector
x_i	The i-th feature value
Φ^{HFM}	The result of HFM model
Φ^{GMF}	The result of GMF model
P_u	The latent feature of users
P_i	The latent feature of items
ω_0	The bias of linear model
ω_i	The weights of linear model
W	The weights of RBM model
$bias_a$	The bias of RBM visual layer
$bias_b$	The bias of RBM hidden layer
$\odot$	Inner product
$\circledast$	HRRs operation

General Matrix Factorization (GMF) [7] is a model that uses a neural network to implement the matrix factorization method. The inputs of this model are the user vector and the item vector. Afterward, the results are concatenated and fed into an MLP to generate the recommendation score.

Factorization machines (FMs) [16] are generic supervised learning models that map arbitrary real-valued features into a low-dimensional latent factor space and can be applied naturally to a wide variety of prediction tasks including regression, classification, and ranking. The FMs model equation is made up of n-way interactions between features. The FMs are defined as:

$$F(x) = \omega_0 + \sum_{i=1}^{n} \omega_i x_i + \sum_{i=1}^{n} \sum_{j=i+1}^{n} <v_i, v_j> x_i x_j \qquad (1)$$

where v_i and x_i represent the i-th dimension latent vector of the features and the i-th feature value respectively, and $< .,. >$ indicates the inner product; $\omega_0 \in \mathbb{R}$ and $\omega_i \in \mathbb{R}^n$ are the parameters of the linear part of FMs model.

Holographic Factorization Machines (HFM) [25] replaced the inner product with HRRs. There are several ways to compute HRRs operations. The most naive way is to literally compress the outer product.

Fortunately, we can exploit computation in the frequency domain, exploiting Fast Fourier Transforms (FFT) for computation with a log-linear runtime. The HHRs operation ⊛ is shown as Eq. 2:

$$a \circledast b = \xi^{-1}(\xi(a) \odot \xi(b)), \qquad (2)$$

where ξ and ξ^{-1} are the fast fourier transforms and inverse fast fourier transforms respectively.

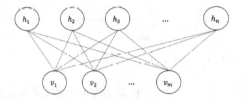

Fig. 1. The RBM structure

Restricted Boltzmann Machines (RBM) [21] can also be used for collaborative filtering. The RBM model structure is shown in Fig. 1. A Restricted Boltzmann Machine is a specific type of Boltzmann machine, which has two layers of units. As illustrated, the first layer consists of visible units, and the second layer includes hidden units. In this restricted architecture, there are no interconnections between units in a single layer.

3   Our HFM++ Model

In this section, we present the detail of our model. As described in earlier sections, to enhance the expressiveness of MF and avoid the disadvantage of the inner product, we propose the HFM++ model. The architecture is shown in Fig. 2.

3.1   Model Architecture

Each training instance of HFM++ accepts an interaction tuple (u, i, r), where u and i indicate user and item respectively. u and i are passed into the network as sparse one-hot-encoded vectors. r represents the rating of the corresponding user and item and it is a binary value with 1 indicating that user u likes item i and 0 indicating dislikes.

Firstly, our proposed model HFM++ concatenates the result of GMF and HFM. The result of GMF and HFM are shown in Eq. 3 and Eq. 4:

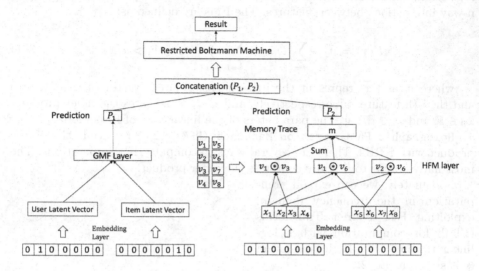

Fig. 2. Architecture of the proposed HFM++ model.

$$\varPhi^{HFM} = L(x) + h^T(\sum_{i=1}^{n}\sum_{j=i+1}^{n}(v_i \circledast v_j)v_iv_j), \qquad (3)$$

$$\varPhi^{GMF} = P_u \odot P_i, \qquad (4)$$

where v_i represents the i-th dimension latent vector of the features, and $\circledast$ refers to HHRs operation; $L(x) = w_0 + \sum_{i=1}^{n} w_ix_i$ represents the linear regression part of the FM formulation, x_i indicates the i-th feature vaule; v_i indicates the i-th dimension latent vector of the features, w_0 and w_i are the parameters of the linear part; and $h \in \mathbb{R}^n$. P_u and P_i are the latent feature of users and items, respectively.

Then, we put the contatenation result into an RBM model, to capture the non-linear latent feature. The RBM model has a hidden layer and a visual layer, whose roles and relationship are shown in Eq. 5 and Eq. 6:

$$hidden\_res = W * \begin{bmatrix} \varPhi^{GMF} \\ \varPhi^{HFM} \end{bmatrix} + bias_b, \qquad (5)$$

$$visual\_res = W * hidden\_res + bias_a, \qquad (6)$$

where *hidden_res* and *visual_res* are the results of hidden layer and visual layer, and $\begin{bmatrix} \Phi^{GMF} \\ \Phi_{HFM} \end{bmatrix}$ indicate the combination result of HFM and GMF model, W is the weight of hidden layer and visual layer. $bias_a$ and $bias_b$ is the bias of the visual layer and hidden layer respectively.

As we can see in Fig. 2, there are three parts in our model architecture, including GMF, HFM, and RBM.

Embedding Layer. The input data passes through an Embedding layer. The input layer is a fully connected layer that projects the one-hot encoded sparse representation into a dense real-valued representation. The user data and item data in the recommendation dataset can be seen as the latent vector for the user data and item data in the context of the latent factor model. The Embedding layer simply transforms each integer i into the ith line of the embedding weights matrix. As such, this layer is parameterized by $W_p \in \mathbb{R}^{|P| \times k}$ and $W_q \in \mathbb{R}^{|Q| \times k}$ respectively. P is the set of users and Q is the set of items. After embedding layer, the user embedding and item embedding are fed into neural network architecture, which can map the latent vectors to prediction scores.

HFM/GMF Layer. After the embedding layer, there is an HFM layer in the HFM part. This layer is mainly used to model the interaction between user and item embeddings. We concatenate the user and item embedding. The HFM models pairwise interactions between each latent feature which have been described in earlier sections. The output of this layer is a scalar value, which represents the score of the user-item pair. HFM model enhanced representation capacity due to compression of outer products and associative retrieval mechanisms.

As for the GMF part, after feeding the input data to the embedding layer, we can get the users and items latent vector. Let the user latent vector p_u and the item latent vector q_i, then we do the element-wise product process for p_u and q_i ($p_u \odot q_i$), and the result is fed to an activation function, which projects the vector to the output layer. The GMF model generalizes the normal MF model to a non-linear setting. It leads to a stronger expressive ability than the linear MF model.

RBM Layer. After getting the predictions of GMF and HFM, we concatenate these two predictions and feed them to a Restricted Boltzmann Machine (RBM) model. Finish training GMF, HFM and RBM part, we get the results of the recommendation.

3.2   Model Learning

Our model has then trained the GMF and HFM parts by using stochastic gradient descent (Adam optimizer), minimizing the binary cross-entropy loss between the targets and the predicted values. We scale the output of GMF and HFM between $[0, 1]$ by using a sigmoid function. The cross-entropy loss is shown as follows:

$$J(\theta) = -y(log(p)) + (1 - y)log(1 - p) + \lambda \|\theta\|_{L2} \tag{7}$$

where y is the ground truth label, $p = \sigma(F(x))$ is the output of the model and is the $\|\theta\|_{L2}$ regularization term weighted by λ.

The algorithm of HFM++ is shown in Algorithm 1. The HFM model can compress outer products and associative retrieval, while the GMF model can get the non-linear feature which has a stronger expressive ability than the linear MF model. As a result, we infer that the combination of HFM and GMF can perform better than the baseline model. On this basis, we fed the combination of HFM and GMF to an RBM model due to the advantage of RBM, which is strong representability and easy to reason. In a word, based on the analysis in this section, we made the hypothesis that this model would perform well, and we confirmed our hypothesis by experiments.

Algorithm 1. HFM++

Input: a batch of input data - x
Output: binary values (1: recommend to the user; 0: do not recommend to the user)
1: $num\_field \leftarrow x.shape[1]$ /*get the second dimension of input data x*/
2: $copy\_x \leftarrow x$ /*get the copy of input data x*/
3: $row \leftarrow emptyList$
4: $col \leftarrow emptyList$
5: **for** $i \leftarrow 0$ to num_field-1 **do**
6:     **for** $j \leftarrow i + 1$ to num_field **do**
7:         row.add (i)
8:         col.add (j)
9:     **end**
10: **end**
11: $ccov\_res \leftarrow sum(x[:, row], x[:, col], dim = 2)$
12: $hfm\_linear\_layer \leftarrow FeaturesLinear((x.shape[0], x.shape[1]), 1)$ /*get the linear layer result by pytorch library*/
13: $\Phi(HFM) \leftarrow hfm\_linear\_layer + ccov\_res$
14: $user\_id \leftarrow copy\_x.shape[0]$
15: $item\_id \leftarrow copy\_x.shape[1]$
16: $\Phi(GMF) \leftarrow FeaturesLinear(user\_id * item\_id)$
17: $input \leftarrow concatenate(\Phi(GMF), \Phi(HFM))$
18: $W \leftarrow randn()$ $a \leftarrow randn()$ $b \leftarrow randn()$ /*parameters of RBM*/
19: **for** $k \leftarrow 1$ to 10 **do**
20:     $hidden\_res \leftarrow W * input + b$
21:     $visual\_res \leftarrow W * hidden\_res + a$
22:     $input \leftarrow visual\_res$
23: predict_result $= visual\_res$

Table 2. Statistics of four datasets

Dataset	Users	Items	Rating
Movielen 20M	7K	2.8K	0.6M
Movielen 10M	7K	2.8K	0.6M
Movielen 1M	6K	4K	1M
Movielen 100K	1K	0.9K	100K

4 Empirical Evaluation

In Sect. 4, we describe our experiments and present empirical results. Firstly, we define the research questions (RQ) that our experiments are designed to answer.

- **RQ1** Does the HFM++ model achieve state-of-the-art performance on CF benchmarks?
- **RQ2** Does the combination improve the performance on FMs and AFM?
- **RQ3** What are the impacts of some key hyperparameters (latent dimensions, learning rate) on performance?

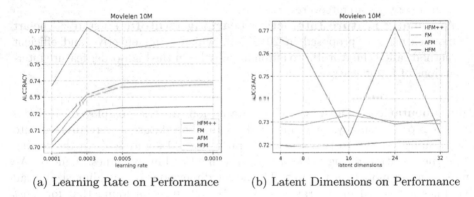

(a) Learning Rate on Performance   (b) Latent Dimensions on Performance

Fig. 3. Effect of hyperparameter on accuracy performance

4.1 Experimental Setup

Datasets. In our experiment, we compare our model against others using four sizes of Movielens[1] dataset. MovieLens is a popular benchmark for recommendation system [1,10]. This dataset is concerned with movie ratings. We utilize four different sizes of this dataset which have 20M (MovieLens20M), 10M (Movie-Lens10M), 1M (MovieLens1M) and 100K (MovieLens100K) respectively where the suffix denotes the number of interactions in the dataset. Due to the hardware

---

[1] https://grouplens.org/datasets/movielens/.

limitation, we just use the parts of the Movielen20M and Movielens10M dataset, which contain 7000 users and 2800 items, due to the limitation of our computer resources. The statistics of datasets we used are as Table 2

Baselines. We compare with one standard baseline and two state-of-the-art models as follows:

- **Factorization Machines (FM)** is a strong CF baseline. It learns pairwise feature interactions using factorized parameters. We concatenate the user and item embedding as input into a standard FM model.
- **Holographic Factorization Machines (HFM)** is a state-of-the-art model proposed by Yi Tay et al. [25]. This model replaces the inner product with HRRs in the FM model.
- **Attentional Factorization Machines (AFM)** is a state-of-the-art model proposed by Xiao et al. [26]. This model follows the implementation of our FM model. However, an attention mechanism is applied on top of FM enabling it to select the best and most informative pairwise features to be used for prediction.
- **Deep Factorization Machine (deepFM)** is a state-of-the-art model proposed by Huifeng Guo et al. [6]. DeepFM combines the power of factorization machines for recommendation and deep learning for feature learning in a new neural network architecture.
- **Automatic Feature Interaction Learning (AutoInt)** a state-of-the-art model that was proposed in 2019 [23], which is an effective and efficient method called the AutoInt to automatically learn the high-order feature interactions of input features.

Implementation Details. The evaluation metric employed includes roc_auc_score, accuracy, precision, and recall. The datasets were divided into the training set and testing set, whose ratios are 80% and 20%. In the testing step, the resulting rating values are transformed as binary values: 0 and 1. We implement all models in Pytorch. The latent dimensions (embedding size) of all baselines are tuned in the range of 4, 8, 16, 32 since we found that for most datasets, performance does not increase beyond $k = 32$. The batch size is set to 1024 in all our experiments. All methods are optimized with Adam with a learning rate of 0.0003. All models are optimized with sigmoid cross-entropy loss since we found that it was significantly more stable compared to minimizing the raw mean squared error. Therefore, rating values are scaled to 0 and 1 and then renormalized upon inference. The number of nodes of the RBM hidden layer is set as 300. We train each model for a maximum of 50 epochs and compute the score on the held-out set at every epoch.

4.2   Experimental Results

RQ1: Does the IIFM++ Model Achieve State-of-the-art Performance on CF Benchmarks? Table 3 reports the overall performance comparison of all compared models and baselines in four metrics. Firstly, we note that HFM++ consistently achieves state-of-the-art performance, outperforming strong baselines on all these four datasets against almost all four metrics. Amongst the baselines, we find that the model that we proposed generally outperforms FM, AFM, HFM, deepFM and AutoInt. The performance of our model in accuracy, precision, and recall metrics outperform all other three models in all four benchmarks.

RQ2: Does the Combination Improve the Performance on FMs and AFM? Table 4 reports the relative performance improvement of our proposed model and FM, AFM respectively. We observe that, in most cases, our proposed model is superior to FM, AFM and AutoInt. The improvements we get are from 1% to 12%. The performances in movielens 100k of recall metric are significantly improved, which are +12.03% and +11.33%.

RQ3: What Are the Impacts of Some Key Hyperparameters (latent Dimensions, Learning Rate) on Performance? Figure 3 shows the effect

Table 3. Performance comparison all metrics of all models on benchmark datasets

Dataset	roc_auc_score					
	FM	IIFM	AFM	deepFM	AutoInt	HFM++
Movielen 20M	**0.7196**	0.7101	0.7177	0.7133	0.7125	0.7176
Movielen 10M	0.7178	0.7082	0.7168	0.735	0.7366	0.7253
Movielen 1M	0.7164	0.7109	0.7217	0.7179	0.719	**0.7536**
Movielen 100k	0.6644	0.6711	0.6729	0.6935	**0.6947**	0.6704

Dataset	Accuracy					
	FM	HFM	AFM	deepFM	AutoInt	HFM++
Movielen 20M	0.7352	0.7306	0.7380	0.7403	0.7408	**0.7696**
Movielen 10M	0.7323	0.7199	0.7321	0.7517	0.7515	**0.7520**
Movielen 1M	0.7323	0.7253	0.7363	0.7325	0.7316	**0.7834**
Movielen 100k	0.6755	0.6708	0.6764	**0.7072**	0.7081	0.69284

Dataset	Precision					
	FM	HFM	AFM	deepFM	AutoInt	HFM++
Movielen 20M	0.7555	0.7541	0.7531	0.7628	0.7609	**0.7880**
Movielen 10M	0.7447	0.7415	0.7438	0.7675	0.7716	**0.7811**
Movielen 1M	0.7500	0.7459	0.7454	0.7441	0.7486	**0.8067**
Movielen 100k	0.6899	0.6925	0.6800	0.6995	0.7014	**0.7395**

Dataset	Recall					
	FM	HFM	AFM	deepFM	AutoInt	HFM++
Movielen 20M	0.8224	0.8081	0.8296	0.8346	0.8399	**0.8544**
Movielen 10M	0.8058	0.7902	0.8168	0.8293	0.8204	**0.8373**
Movielen 1M	0.8161	0.8040	0.8248	0.8149	0.8027	**0.8631**
Movielen 100k	0.7509	0.7693	0.7556	0.8244	0.8216	**0.8412**

of varying the learning rates and latent dimension for all models on Movielens 10M dataset. What we can see from the Fig. 3(a) is that the HFM++ model clearly outperforms all three other models on all learning rates. HFM++ gets the best performance when the learning rate is 0.0003. Figure 3(b) reports the effect of different latent dimensions k for all four models on Movielens 10M. For HFM++, the optimal dimensions are k = 24 on Movielens 10M. Also, we can see that there are 3 latent dimensions (4, 8, and 24) that achieve much better performances than FM, AFM, and HFM models.

5 Related Work

The term Collaborative Filtering was first proposed by [5], which is the first recommender system. A lot of cutting-edge research in this field applied a factorization-based approach to learning the relationship between users and items. In particular, probabilistic matrix factorization [13] and SVD [9] were amongst the more popular CF algorithms. Factorization machines [16,17] were later proposed as a general-purpose machine learning model, i.e., regression and classification. Moreover, Factorization Machines also see wide adoption in the recommendation system field [8,14,18,29]. The principle of General Matrix Factorization (GMF) [3] is also based on the factorization to get the relationship between users and items, however, the GMF model uses a neural network to learn the user-item associations instead of just matrix factorization.

Owing to its effectiveness in capturing feature interactions, FMs have also been recently adopted in other variations of recommender tasks such as review-

Table 4. Relative performance improvement (+%) against FM, AFM, HFM, deepFM and AutoInt model

Dataset	HFM++ vs FM			
	ROC_AUC_SCORE	ACCURACY	PRECISION	RECALL
Movielen 20M	−0.28%	+4.68%	+4.30%	+3.89%
Movielen 10M	+1.04%	+2.69%	+4.89%	+3.91%
Movielen 1M	+5.19%	+6.98%	+7.56%	+5.76%
Movielen 100k	+0.90%	+2.56%	+7.19%	+12.03%
Dataset	HFM++ vs AFM			
	ROC_AUC_SCORE	ACCURACY	PRECISION	RECALL
Movielen 20M	−0.01%	+4.28%	+4.63%	+2.99%
Movielen 10M	+1.19%	+2.72%	+5.01%	+2.51%
Movielen 1M	+4.42%	+6.40%	+8.22%	+4.64%
Movielen 100k	−0.37%	+2.42%	+8.75%	+11.33%
Dataset	HFM++ vs AutoInt			
	ROC_AUC_SCORE	ACCURACY	PRECISION	RECALL
Movielen 20M	+0.72%	+3.89%	+3.56%	+1.73%
Movielen 10M	−1.53%	+0.07%	+1.23%	+2.06%
Movielen 1M	+4.81%	+7.08%	+7.76%	+5.20%
Movielen 100k	−3.50%	−2.16%	+5.43%	+2.54%

based or sequential recommender systems. DeepCoNN [29] used an FM on top of user and item convolutional neural network (CNN) for review rating prediction. The recent Multi-Pointer Co-Attention Networks [24] similarly adopts an FM prediction layer. [14] proposed a translational sequential recommender model based on factorization machines.

HFM [25] is method that inspired by applications of HRR [15] and associative memory models [4]. HFM replaces the inner product of FM with HRR and achieves state-of-the-art recommendation tasks.

6   Conclusion

In this paper, we proposed a novel model named HFM++ that integrates an RBM and a GMF into HFM, resulting in a more powerful recommendation model that inherits the benefits of all these three models. In the experiment phase, we used four benchmark datasets including Movielens100k, Movielens1M, Movielens10M, and Movielens20M, and four metrics including roc_auc_score, accuracy, precision, and recall. We compared our HFM++ model against five baseline models (FM, AFM, HFM, deepFM and AutoInt), and we got the result that our proposed model outperforms the other five models in all these four benchmark datasets with the metric of accuracy, precision, and recall. HFM++ achieves comparable or better performances against the strongest baselines with a performance boost up to 12%. In the future, we will explore more effective operators to enhance its capability in interaction modeling.

References

1. Bai, R., Zhu, H., Ni, Y., Chen, Y., Zheng, Q.: Autocot-autoencoder based cooperative training for sparse recommendation. In: 2018 IEEE 15th International Conference on e-Business Engineering (ICEBE), pp. 257–262. IEEE (2018)
2. Chen, H., Shi, S., Li, Y., Zhang, Y.: Neural collaborative reasoning. In: Proceedings of the Web Conference 2021, pp. 1516–1527 (2021)
3. Chen, T., Li, H., Yang, Q., Yu, Y.: General functional matrix factorization using gradient boosting. In: International Conference on Machine Learning, pp. 436–444. PMLR (2013)
4. Gábor, D.: Associative holographic memories. IBM J. Res. Dev. **13**(2), 156–159 (1969)
5. Goldberg, D., Nichols, D., Oki, B.M., Terry, D.: Using collaborative filtering to weave an information tapestry. Commun. ACM **35**(12), 61–70 (1992)
6. Guo, H., Tang, R., Ye, Y., Li, Z., He, X.: Deepfm: a factorization-machine based neural network for CTR prediction (2017). arXiv preprint arXiv:1703.04247
7. He, X., Liao, L., Zhang, H., Nie, L., Hu, X., Chua, T.S.: Neural collaborative filtering. In: Proceedings of the 26th International Conference on World Wide Web, pp. 173–182 (2017)
8. Juan, Y., Zhuang, Y., Chin, W.S., Lin, C.J.: Field-aware factorization machines for CTR prediction. In: Proceedings of the 10th ACM Conference on Recommender Systems, pp. 43–50 (2016)

9. Koren, Y.: Factorization meets the neighborhood: a multifaceted collaborative filtering model. In: Proceedings of the 14th ACM SIGKDD International Conference on Knowledge Discovery and Data Mining, pp. 426–434 (2008)
10. Lin, K., Li, B.: Collaborative and content-based filtering with data from movielen. Int. Core J. Eng. **7**(4), 28–30 (2021)
11. Lu, J., Wu, D., Mao, M., Wang, W., Zhang, G.: Recommender system application developments: a survey. Decis. Supp. Syst. **74**, 12–32 (2015)
12. Lü, L., Medo, M., Yeung, C.H., Zhang, Y.C., Zhang, Z.K., Zhou, T.: Recommender systems. Phys. Rep. **519**(1), 1–49 (2012)
13. Mnih, A., Salakhutdinov, R.R.: Probabilistic matrix factorization. Adv. Neural Inf. Process. Syst. **20**, 1257–1264 (2007)
14. Pasricha, R., McAuley, J.: Translation-based factorization machines for sequential recommendation. In: Proceedings of the 12th ACM Conference on Recommender Systems, pp. 63–71 (2018)
15. Plate, T.A.: Holographic reduced representations. IEEE Trans. Neural Netw. **6**(3), 623–641 (1995)
16. Rendle, S.: Factorization machines. In: 2010 IEEE International Conference on Data Mining, pp. 995–1000. IEEE (2010)
17. Rendle, S.: Factorization machines with libfm. ACM Trans. Intell. Syst. Technol. (TIST) **3**(3), 1–22 (2012)
18. Rendle, S., Gantner, Z., Freudenthaler, C., Schmidt-Thieme, L.: Fast context-aware recommendations with factorization machines. In: Proceedings of the 34th International ACM SIGIR Conference on Research and Development in Information Retrieval, pp. 635–644 (2011)
19. Resnick, P., Varian, H.R.: Recommender systems. Commun. ACM **40**(3), 56–58 (1997)
20. Ricci, F., Rokach, L., Shapira, B.: Introduction to recommender systems handbook. In: Ricci, F., Rokach, L., Shapira, B., Kantor, P.B. (eds.) Recommender Systems Handbook, pp. 1–35. Springer, Boston, MA (2011). https://doi.org/10.1007/978-0-387-85820-3_1
21. Salakhutdinov, R., Mnih, A., Hinton, G.: Restricted Boltzmann Machines for collaborative filtering. In: Proceedings of the 24th International Conference on Machine Learning, pp. 791–798 (2007)
22. Schafer, J.B., Konstan, J., Riedl, J.: Recommender systems in e-commerce. In: Proceedings of the 1st ACM Conference on Electronic Commerce, pp. 158–166 (1999)
23. Song, W., et al.: Autoint: Automatic feature interaction learning via self-attentive neural networks. In: Proceedings of the 28th ACM International Conference on Information and Knowledge Management, pp. 1161–1170 (2019)
24. Tay, Y., Luu, A.T., Hui, S.C.: Multi-pointer co-attention networks for recommendation. In: Proceedings of the 24th ACM SIGKDD International Conference on Knowledge Discovery & Data Mining, pp. 2309–2318 (2018)
25. Tay, Y., Zhang, S., Luu, A.T., Hui, S.C., Yao, L., Vinh, T.D.Q.: Holographic factorization machines for recommendation. In: Proceedings of the AAAI Conference on Artificial Intelligence, vol. 33, pp. 5143–5150 (2019)
26. Xiao, J., Ye, H., He, X., Zhang, H., Wu, F., Chua, T.S.: Attentional factorization machines: Learning the weight of feature interactions via attention networks (2017). arXiv preprint arXiv:1708.04617
27. Yao, Q., Wang, Y., Kwok, J.T.: A scalable, adaptive and sound nonconvex regularizer for low-rank matrix completion. arXiv e-prints pp. arXiv-2008 (2020)

28. Zhang, Y., Zhu, Z., He, Y., Caverlee, J.: Content-collaborative disentanglement representation learning for enhanced recommendation. In: Fourteenth ACM Conference on Recommender Systems, pp. 43–52 (2020)
29. Zheng, L., Noroozi, V., Yu, P.S.: Joint deep modeling of users and items using reviews for recommendation. In: Proceedings of the tenth ACM International Conference on Web Search and Data Mining, pp. 425–434 (2017)

Deep Learning for Bias Detection: From Inception to Deployment

Md Abul Bashar[1]([✉])([iD]), Richi Nayak[1], Anjor Kothare[2], Vishal Sharma[2], and Kesavan Kandadai[2]

[1] Queensland University of Technology, Brisbane, Australia
{m1.bashar,r.nayak}@qut.edu.au
[2] ishield.ai, San Francisco, CA, USA
{anjor.kothare,kesavan}@ishield.ai, vishal.sharma@tothenew.com,
kesavan@pallavatech.com

Abstract. To create a more inclusive workplace, enterprises are actively investing in identifying and eliminating unconscious bias (e.g., gender, race, age, disability, elitism and religion) across their various functions. We propose a deep learning model with a transfer learning based language model to learn from manually tagged documents for automatically identifying bias in enterprise content. We first pretrain a deep learning-based language-model using Wikipedia, then fine tune the model with a large unlabelled data set related with various types of enterprise content. Finally, a linear layer followed by softmax layer is added at the end of the language model and the model is trained on a labelled bias dataset consisting of enterprise content. The trained model is thoroughly evaluated on independent datasets to ensure a general application. We present the proposed method and its deployment detail in a real-world application.

Keywords: Deep learning · Bias detection · Transfer learning · Text data · Small dataset

1 Introduction

A rigorous study by McKinsey [18] found that more diverse and inclusive organisations outperform those that are not. The concept of unconscious bias has become increasingly pervasive, with many organisations training their employees on the concept of Diversity and Inclusion (D&I). Unfortunately, though an increased awareness of unconscious bias can have benefits, it is not a systemic and consistent solution [24]. Enterprises are looking for technologies that can create consistent and scalable practices to identify or mitigate biases across organisations, often in real-time.

While there have been a few analytics products to measure employee demographics, pay parity and customised training [39], there is lack of solutions targeted toward tackling unconscious bias in content that is created within an enterprise. For every enterprise, almost always the first touch point for their

© Springer Nature Singapore Pte Ltd. 2021
Y. Xu et al. (Eds.): AusDM 2021, CCIS 1504, pp. 86–101, 2021.
https://doi.org/10.1007/978-981-16-8531-6_7

users, vendors, employees or business partners is the content published across their digital channels. This content may be job descriptions, website content, marketing messages, blogs, reports and social media content. Internally within the enterprise, content is also created in the form of product documentation, user guides, customer care, messaging and collaboration platforms. Unconscious bias in content manifests itself in various forms in different types of content. For example: (a) in job descriptions this could be through use of pronouns or masculine coded words, (b) in enterprise messaging platforms this could be through toxic messages and bullying, (c) in product descriptions this could be through stereotyping a type of a user, (d) in e-commerce product details this could be through references to body shapes, or (e) in a stock market report, this could be through inherent assumption of the gender of a potential investor.

In this paper, we aim to propose a Deep Learning (DL) model to detect unconscious bias in content and how it can be applied to integrate and work with enterprise applications. Lexical based systems and traditional machine learning systems do not solve the problem due to complexity and intricate relationships inherent in the textual narratives. With advancements in DL in Natural Language Processing (NLP), it is feasible to build a DL based system. However, such a system requires a large set of labelled data for building the model. With the manual efforts involved in labelling the data, it is difficult to create a large dataset. There do not exist similar data that can be used in labelling. In this paper we propose a novel method based on transfer learning to deal with the small set of labelled data and detect unconscious bias in content.

This research makes the following novel contributions. (1) It proposes a comprehensive bias detection model that can detect four types of bias (*Race, Gender, Age, Not Appropriate*) in text data. (2) It uses progressive transfer learning that allows to train a smaller model (i.e. less number of hyper parameters) with a small labelled dataset for better accuracy. (3). It presents the detail of deploying the model in a real-world application[1].

2   Literature Review

When the unconscious biases are not tackled, they cause serious harm to the business including financially, socially and culturally [1,3,6,18,28,46]. However, very few to no technology solutions exist to tackle unconscious bias in the content that is created and published across the enterprise for both their internal and external audiences [20].

Deep learning models have become quite successful in natural language processing, e.g., content generation, language translation, Question and Answering systems, text classification [2,7–10] and clustering. In spite of this success of DL in NLP, there has been very limited works on building a generic bias detection method. The work in current literature can be grouped into two categories: (1) text representation learning; and (2) reducing subjective bias in text.

---

[1] https://ishield.ai/.

Researchers[2] [11,33] proposed to debias semantic representation of words by removing bias component from word embedding. The goal is to make semantic representations fair across attributes like gender and age. Autoencoder was used to generate a balanced gender-oriented word distribution to remove gender bias from word embeddings[30]. Counterfactual data has been augmented to alter the training distribution to balance gender-based statistics [16,34,47]. Research in [44] used adversarial training to squeeze directions of bias in the hidden states of image representation.

Research in [41] proposed a model for automatically suggesting edits for subjective-bias words following the Wikipedia's neutral point of view (NPOV) policy for defining subjective bias. However, the model is limited to single-word edits, i.e. it can handle simplest instances of bias only. Research in [27,42] used lexicon of bias words for detecting language bias in sentences of Wikipedia articles. A fine-tuned BERT model [15] is used for detecting gender bias only [17].

An issue faced by bias detection methods in multi-class setting is failing to select minority class examples in imbalanced data distribution [5]. Guided learning, based on crowd-sourcing to find or generate class-specific training instances, can help to get more balanced class frequencies [4,40]. However, guided learning is resource consuming and may not present the true distribution generating training examples. Sometimes, heuristic labelling methods such as distant supervision [14] or data programming [19] are used for datasets with the imbalanced class distribution. However, these methods are only applicable when a good knowledge base or a pretrained predictor is available [13].

In this research, we propose to use progressive transfer learning to address the class imbalance problem. To our best of knowledge, there exist no model that comprehensively detect common biases (e.g. race, gender, age) in text.

3   Language Based Deep Learning Model for Bias Detection

Bias detection in the text data is a complex problem because usually bias is represented by linguistic cues that are subtle and can be determined only through its context in the text. Let X be a text dataset that contains n features and K classes. Let $\mathbf{x} = \langle x_1, \dots x_n \rangle$ be a vector representing an instance in X. Let C_k be a set of K classes. The bias detection is a classification task that assigns an instance to a bias class (or category) C_k based on the feature vector $\mathbf{x}$; i.e. $f \in \mathcal{F} : X \to C_k$, where $f(\mathbf{x}) = \max_{C_k} p(C_k|\mathbf{x})$. This ascertains that we need to know $p(C_k|\mathbf{x})$ for a bias detection task. The joint probability $p(\mathbf{x}, C_k)$ of $\mathbf{x}$ and C_k can be written as

$$p(\mathbf{x}, C_k) = p(C_k|\mathbf{x})p(\mathbf{x}) \tag{1}$$

where $p(\mathbf{x})$ is the prior probability distribution. The prior probability $p(\mathbf{x})$ can be seen as a regulariser for $p(C_k|\mathbf{x})$ that can regularise modelling of the associated uncertainties of $p(\mathbf{x}, C_k)$ [10]. As $p(\mathbf{x})$ does not depend on C_k, this means that

---

[2] https://gender-decoder.katmatfield.com/.

$p(\mathbf{x})$ can be learned independent of the class level C_k. That is, $p(\mathbf{x})$ can be learned from unlabelled data.

Prior research [9,10] showed that the estimation of $p(\mathbf{x}, C_k)$ can be improved when $p(\mathbf{x})$ is learned from a sequence of unlabelled datasets, especially when the labelled dataset is small. In this study, we propose to implement a bias classification model utilising this technique of improving the prediction accuracy through unlabelled data.

A discriminative model such as LSTM learns to classify an instance $\mathbf{x}$ into class C_k by learning the conditional probability distribution as $p(C_k|\mathbf{x}, \theta) \approx p(C_k|\mathbf{x})$ where θ is the list of model parameters. However, accurately approximating $p(C_k|\mathbf{x})$ requires a large number of labelled instances. If only a small set of labelled data is available, the learned $p(C_k|\mathbf{x}, \theta)$ might not be a good approximation of population distribution because θ may over-fit the small training set. Alternately, $p(\mathbf{x})$ can be learned from one or more large unlabelled datasets and conditioned on $\mathbf{x}$ to learn $p(C_k|\mathbf{x}, \theta)$, leading to $p(C_k|\mathbf{x}, \theta)p(\mathbf{x}) \approx p(\mathbf{x}, C_k)$. The term $p(C_k|\mathbf{x}, \theta)p(\mathbf{x})$ can be seen as equivalent to combining the regularisation into the discriminative model. This regularised model would act similar to a generative model.

Unlike common transfer learning where $p(\mathbf{x})$ is learned once from a large unlabelled dataset, we propose to progressively learn $p(\mathbf{x})$ from a sequence of unlabelled datasets. This allows us to use a relatively small set of parameters in our model. Then we use $p(\mathbf{x})$ to learn $p(C_k|\mathbf{x}, \theta)p(\mathbf{x}) \approx p(\mathbf{x}, C_k)$ with a small training dataset. Next, we present the estimation of $p(\mathbf{x})$ as a neural network language model (NNLM) using unsupervised learning.

3.1   Neural Network Language Model

Probability $p(\mathbf{x})$ can be estimated using the assumption of Language model where features are considered conditionally dependent [9,10]. This is to support natural language processing where in a sentence, the sequencing of words depends on each other. Based on this, the joint probability $p(\mathbf{x}, C_k)$ in Eq. 1 can be rewritten as follows, using the chain rule:

$$
\begin{aligned}
p(\mathbf{x}, C_k) &= p(C_k|\mathbf{x})p(\mathbf{x}) \\
&= p(C_k|\mathbf{x})p(x_1, \ldots, x_n) \\
&= p(C_k|\mathbf{x}) \prod_{i=1}^{n} p(x_i|x_1 \ldots x_{i-1})
\end{aligned}
\tag{2}
$$

The part $\prod_{i=1}^{n} p(x_i|x_1 \ldots x_{i-1})$ in Eq. 2 can be considered as a language model because $p(x_i|x_1 \ldots x_{i-1})$ seeks to predict the probability of observing the ith feature x_i, given the previous $(i-1)$ features $(x_1 \ldots x_{i-1})$. A Recurrent Neural Network (RNN) or its variants such as Long Short-Term Memory (LSTM) can be used to model $\prod_{i=1}^{n} p(x_i|x_1 \ldots x_{i-1})$ as they work in a similar way to capture the order of features and their non-linear and hierarchical interactions [10, 29, 38]. Similar to traditional language models, a RNN/LSTM based NNLM can approximate joint probabilities over the feature sequences as follows, where ω is the list of model parameters.

$$
\begin{aligned}
p(\mathbf{x}) &= p(x_1, \ldots, x_n) \\
&\approx p(x_1, \ldots, x_n, \omega) \\
&\approx \prod_{i=1}^{n} p(x_i | x_1 \ldots x_{i-1}, \omega)
\end{aligned}
\tag{3}
$$

Given a sequence of features, a RNN recurrently processes each feature and uses multiple hidden layers to capture the order of features and their non-linear and hierarchical interactions. The hidden state is used to derive a vector of probabilities representing the network's guess of the subsequent feature in the sequence [9,10]. The network aims to minimize the loss calculated based on the vector of probabilities and the actual next feature. In simple words, the context of all previous features in the sequence is encoded within the parameters ω of the network and the probability of getting the next word is distributed over the vocabulary using a Softmax function [29].

Estimating $p(\mathbf{x})$ using a LSTM-based LM (i.e. NNLM) model on a huge dataset that covers multitude of domains is useful for transfer learning, but it can be very expensive in terms of required computation and memory [12]. Additionally, it can learn irrelevant and misleading relationships in data due to interactions between different domains in a single corpus [10]. Therefore, as suggested in [9], we use an alternative approach based on progressive transfer learning to incorporate knowledge gained from a sequence of datasets in a LSTM-based LM [10].

Let there be m number of corpora from which the knowledge is gained. A LSTM model built on corpus D_i to learn $p(\mathbf{x}|D_i)p(D_i)$ will have its parameters ω_i. It can be expressed as follows.

$$
\begin{aligned}
\prod_{i=1}^{m} p(\mathbf{x}|D_i)p(D_i) &\approx \prod_{i=1}^{m} p(\mathbf{x}|D_i, \omega_i)p(D_i, \omega_i) \\
&= \prod_{i=1}^{m} p(\mathbf{x}|D_i, \omega_i)p(\omega_i|D_i)p(D_i)
\end{aligned}
\tag{4}
$$

If the same LM model is sequentially built from the given m datasets, parameters ω_i learned on i^{th} dataset will only depend on the parameters ω_{i-1} learned on the $(i-1)^{th}$ dataset, applying the Markov assumption.

$$
\prod_{i=1}^{m} p(\mathbf{x}|D_i, \omega_i)p(\omega_i|D_i)p(D_i) \approx \prod_{i=1}^{m} p(\mathbf{x}|D_i, \omega_i)p(\omega_i|D_i, \omega_{i-1})p(D_i)
\tag{5}
$$

Here ω_0 is the initial weight that might be assigned randomly. Assuming the same probability (or uncertainty) for each dataset, transfer learning can be expressed as follows.

$$p(\mathbf{x}, D_1, \ldots, D_n) \approx \prod_{i=1}^{m} p(\mathbf{x}|D_i, \omega_i) p(\omega_i|D_i, \omega_{i-1}) p(D_i)$$

$$= \prod_{i=1}^{m} p(\mathbf{x}|D_i, \omega_i) p(\omega_i|D_i, \omega_{i-1}) \tag{6}$$

$$\propto \sum_{i=1}^{m} ln \left(p(\mathbf{x}|D_i, \omega_i) p(\omega_i|D_i, \omega_{i-1}) \right)$$

Followings can be inferred from Eq. 6. (1) Each dataset D_i relevant to the application domain of LM can reduce uncertainty [8,10]. (2) Pr-training of LSTM-LM should be done by the order of the dataset of general population distribution to the dataset of specific population distribution because the parameter vector ω_i depends on ω_{i-1} [9,10]. For example, we can approximate the population distribution of Queensland (i.e., specific) from that of Australia (i.e., general) but the opposite is not true.

3.2   Regularising Classifier with Language Model

For the downstream task of classification, a LSTM is trained to learn $p(C_k, \mathbf{x})$ $\approx p(C_k|\mathbf{x}, \theta) p(\mathbf{x})$ on a small training dataset. The prior distribution $p(\mathbf{x})$ can be learned by sequentially pretraining an LSTM on m unlabelled datasets from general to specific domain. Using Eq. 6, we can write the regularised classifier as follows.

$$p(C_k, \mathbf{x}) \approx p(C_k, \mathbf{x}) p(\mathbf{x}, \omega)$$

$$\approx p(C_k|\mathbf{x}, \theta) \prod_{i=1}^{m} p(\mathbf{x}|D_i, \omega_i) p(\omega_i|D_i, \omega_{i-1}) \tag{7}$$

Figure 1 shows the process of transfer learning through LSTM-based LM to a LSTM classification model. Layers 1 to 3 are stacked LSTM layers. The LSTM-based LM (the left hand side model in Fig. 1) is generated using three staked layers along with embedding layer and LM softmax layer. The LM softmax layer is active during pretraining of LM with a sequence of unlabelled datasets and then it is frozen. Once the LM is pretrained, the class softmax layer and linear layer are augmented, as shown by the right hand side model in Fig. 1. These two layers along with the pretrained LM active layers are trained with the small labelled dataset to learn the classification task of bias detection.

The main task of these additional two layers is to learn $p(C_k|\mathbf{x}, \theta)$. The combined network learns $p(C_k|\mathbf{x}, \theta) p(\mathbf{x}, \omega)$. The additional two layers are augmented at the end of NNLM in this model to assure that θ is learned from the finetuning of ω with labelled dataset during classification training. This process of training a classifier model (e.g. LSTM) generates a classifier regularised by the language model (LM) based transfer learning [10]. We call the trained model as LSTM-LM.

4   Empirical Analysis

Extensive experiments were conducted to evaluate the accuracy of the proposed method for bias detection. We used six standard classification evaluation measures [10]: Accuracy (Ac), Precision (Pr), Recall (Re), F_1 Score (F_1), Cohen Kappa Score (CKS) and Area Under Curve (AUC).

4.1   Data Collection

The iShield.ai Dataset. (Version 1^3) has a total data count of 57,424 sentences, where 27,131 sentences are biased (47%) and 30,293 sentences are unbiased (53%). There are four different kinds of biases: (1) 20,690 (36%) sentences are GENDER biased, (2) 4,339 (7.56%) sentences are RACE biased, (3) 1,553 (2.7%) sentences are ambiguous and (4) 549 (0.96%) sentences are AGE biased. We used 80–10-10% ratio between training, validation and testing instances. The dataset comes from two sub-domains namely Job Description (JD) and Non Job Description (NJD) that contributed 25,123 (43.75%) and 32,301 (56.25%) sentences respectively.

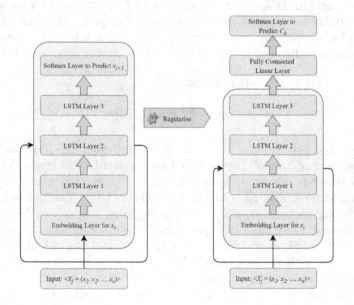

Fig. 1. LSTM-L: Process of regularising a classifier using a language model

We used the following four definitions of bias in this research. **GENDER** – Any conscious or unconscious attempt to single out a particular gender or people who identify with a particular gender identity. Example: The claims assistant

---

[3] In the next versions, this dataset has been considerably modified.

will handles the claims of veterans and their wives. **RACE** – Any conscious or unconscious attempt to single out a particular race or it's people. Example: Own development of brownbag sessions and facilitate publishing reusable content to the organisation. **AGE** – Any conscious or unconscious attempt to single out a particular age bracket or it's people. Example: We are a young organisation looking for young and talented marketers. **Ambiguous** – Any part of a sentence which does not clearly convey the intended meaning. Example: We are looking for a smart candidate for this position.

The basic unit of division used for annotation is a sentence. Any document is first split in sentences. Then sentences are passed to annotators in batches, where each batch consists of 150 sentences. Biases are usually observed as a group of four to five words in a sentence. A three member panel is set up for a Quality Assurance pass. The panel evaluates the labelled sentences for two checks. (1) The labelled sentences are adhering to the outlined definitions of bias. (2) Conflicting instances (e.g. a sentence should have been labelled as biased but is not) are eliminated or placed in the correct category.

Sexist Statement in Workplace (SSW) Dataset [22]. This dataset was collected to check how effectively the classification model works on other datasets besides the bias detection dataset. The SSW dataset has around 1100 labelled instances for sexism statement in workplace. The instances are roughly balanced between sexism (labelled 1) and neutral (labelled 0) cases. Some examples from this dataset is given in Table 1. We used 80–20% ratio between training and testing instances. Hyper parameters were set by using cross validation in the 80% data used for training.

Table 1. Instances from SSW Dataset

Label	Statement
1	Women always get more upset than men
1	The people at work are childish. it's run by women and when womendont agree to something, oh man
1	I'm going to miss her resting bitch face
0	No mountain is high enough for a girl to climb
0	It seems the world is not ready for one of the most powerful andinfluential countries to have a woman leader. So sad
0	Can you explain why what she described there is wrong?

Model Pretraining Datasets. We collected a list of pretrained word vectors from [36]. The list has one million word vectors trained on Wikipedia 2017, UMBC webbase corpus and statmt.org news dataset (16B tokens). We use the following three corpora (D_1, D_2, and D_3) for sequentially pretraining LSTM-LM model and fineturning to target task, i.e. first LSTM-LM is pretrained using D_1, then using D_2 and finally fintuned to target task using D_3.

D_1: The goal of using this corpus is to capture general properties of the English language. We pretrain the LSTM-LM model on Wikitext-103 that contains 28,595 preprocessed Wikipedia articles and 103 million words [35]. After pretraining on D_1, we approximate the probability distribution $p(\mathbf{x}|D_1, \omega_1)$.

D_2: The goal of using this corpus is to bridge the data distribution between the target task domain (i.e., bias detection in job description) and the general domain (i.e. standard language). This is because the target task is likely to come from a different distribution than the general corpus. D_2 should be chosen such that it has commonalities with both D_1 reflecting a general domain (Wikipedia) and the corpus D_3 reflecting a target domain (e.g. labelled data of job description). We use a set of unlabelled JD and NJD data as D_2.

4.2   Baseline Models

We have implemented 10 baseline models to compare the performance of the proposed LSTM-LM.

Models with pretrained word vectors (i.e. word embeddings by word2Vec) include (1) LSTM with pretrained Word vectors (LSTM-W) [25] and (2) CNN with pretrained Word vectors (CNN-W) [8]. LSTM-W has 100 units, 50% dropout, binary cross-entropy loss function, Adam optimiser and sigmoid activation. The hyper parameters of CNN-W is set as in [8]. We used one million pretrained word vectors each with 300-dimension [36]. Word vectors are pretrained on Wikipedia 2017, UMBC webbase corpus and statmt.org news dataset. A Continuous Bag-of-Words Word2vec [37] model is used in pretraining.

LSTM and CNN without pretrained word vectors (LSTM-P) [8,25]. These are traditional LSTM and CNN models that have not been pretrained by any data. Similar to LSTM-W, LSTM-P has 100 units, 50% dropout, binary cross-entropy loss function, Adam optimiser and sigmoid activation. The hyper parameters of CNN is set as in [8].

Feedforward Deep Neural Network (DNN) [21]. It has five hidden layers, each layer containing eighty units, 50% dropout applied to the input layer and the first two hidden layers, softmax activation and 0.04 learning rate. For all neural network based models, hyperparameters are manually tuned based on cross-validation.

Non NN models include Support Vector Machines (SVM) [23], Random Forest (RF) [32], Decision Tree (DT) [43], Gaussian Naive Bayes (GNB) [31], k-Nearest Neighbours (kNN) [45] and Ridge Classifier (RC) [26]. Hyperparameters of all these models are automatically tuned using ten-fold cross-validation and GridSearch using sklearn library.

None of the models, except LSTM-LM, LSTM-W and CNN-W, are pretrained or utilised unlabelled dataset.

4.3   Experimental Results: SSW Dataset

SSW is a very small dataset. The experimental results on SSW dataset are given in Table 2. The proposed language model-based transfer learning model LSTM-LM outperforms all the baseline models. Beside LSTM-LM, other two word vector-based transfer learning-based models, CNN-W and LSTM-W yield the second and third best performance, respectively. We conjecture that (1) transfer learning-based models produce better outcome, and (2) the language model-based transfer learning brings more benefits than the word vector-based transfer learning when the training dataset is small.

It is interesting to note that CNN-W outperforms CNN as well as LSTM-W outperforms LSTM. CNN-W and CNN (or LSTM-W and LSTM) use the exactly same architecture, except that CNN-W (or LSTM-W) uses pretrained word vectors for transfer learning. This further emphasises the benefit of using transfer learning over the standard models when the training dataset is small.

Traditional models (i.e. RF, DT, GNB and kNN) do not utilise any transfer learning and solely rely on the labelled training dataset. Therefore, they give lower performance when the labelled training dataset is small.

4.4   Experimental Results: the iShield.ai Dataset

The experimental results comparing our LSTM-LM against ten baseline models on the iShield dataset are given in Table 3.

Table 2. Experimental results on SSW dataset

	Sample average			Weighted average			Macro average			
	Accu.	AUC	CKS	Prec.	Recall	F₁	Prec.	Recall	F₁	Support
LSTM-LM	**0.89**	**0.89**	**0.77**	**0.89**	**0.89**	**0.89**	**0.89**	**0.88**	**0.89**	228
RF	0.82	0.81	0.63	0.82	0.82	0.82	0.81	0.81	0.81	228
DT	0.79	0.78	0.57	0.79	0.79	0.79	0.78	0.78	0.78	228
GNB	0.70	0.71	0.40	0.72	0.7	0.7	0.71	0.71	0.7	228
kNN	0.58	0.60	0.19	0.63	0.58	0.56	0.62	0.6	0.57	228
SVM	0.82	0.81	0.63	0.82	0.82	0.82	0.82	0.81	0.81	228
RC	0.77	0.77	0.53	0.77	0.77	0.77	0.77	0.77	0.77	228
CNN-W	0.86	0.86	0.72	0.86	0.86	0.86	0.86	0.86	0.86	228
CNN	0.82	0.81	0.63	0.82	0.82	0.82	0.82	0.81	0.82	228
LSTM-W	0.85	0.84	0.70	0.85	0.85	0.85	0.86	0.84	0.85	228
LSTM	0.82	0.82	0.64	0.82	0.82	0.82	0.82	0.82	0.82	228

Accuracy, AUC and CKS are three important measures for understanding the overall significance of a classification model. Table 3 shows that LSTM-LM gives the best Accuracy, AUC and CKS results. CKS indicates the reliability between the prediction made by a model and the ground truth. All the three transfer learning-based models (i.e. LSTM-LM, LSTM-W and CNN-W) have high CKS value. However, LSTM-LM has better accuracy and AUC than other two (i.e. LSTM-W and CNN-W).

Overall LSTM-LM gives us the best results, as indicated by best weighted average precision, recall and F_1 score. Even though SVM, CNN-W and LSTM-W have the same weighted average precision value as LSTM-W, their recall and F_1 score are lower than LSTM-W.

High precision indicates most of the identified biased sentences are indeed bias. However, if the recall is not high enough, then many bias sentences will left undetected. Therefore, better recall is desirable in bias detection. Yet the excessive false positives can result in a higher cost for investigating many false detection. Therefore, a balance in both recall and precision is needed. A higher F_1 score indicates both precision and recall is high. LSTM-LM gives the best F_1 score for both weighted average and macro average. Macro Average is used to evaluate the performance of a classifier for minority classes (classes with small number of instances), where weighted average favours majority classes. The high F_1-score for both weighted average and macro average indicates that LSTM-LM works reasonably well for both majority and minority classes. Best macro average precision is achieved by SVM, but SVM has very poor recall value and F_1-score.

Table 3. Experimental results on the iShield.ai dataset

	Sample average			Weighted average			Macro Average			
	Accu.	AUC	CKS	Prec.	Recall	F_1	Prec.	Recall	F_1	Support
LSTM-LM	**0.85**	**0.78**	**0.73**	**0.84**	**0.85**	**0.84**	0.69	**0.6**	**0.63**	5743
RF	0.84	0.76	0.72	0.83	0.84	0.83	0.72	0.57	0.61	5743
DT	0.83	0.77	0.71	0.83	0.83	0.83	0.64	0.59	0.61	5743
GNB	0.62	0.74	0.45	0.79	0.62	0.67	0.44	0.58	0.43	5743
kNN	0.79	0.70	0.64	0.78	0.79	0.78	0.66	0.48	0.52	5743
SVM	0.84	0.73	0.72	**0.84**	0.84	0.83	**0.81**	0.51	0.55	5743
RC	0.83	0.75	0.70	0.82	0.83	0.82	0.68	0.57	0.6	5743
CNN-W	0.84	0.76	**0.73**	**0.84**	0.84	0.83	0.73	0.57	0.59	5743
CNN	0.84	0.75	0.72	0.83	0.84	0.83	0.58	0.56	0.56	5743
LSTM-W	0.84	0.77	**0.73**	**0.84**	0.84	0.84	0.67	0.58	0.6	5743
LSTM	0.84	0.76	0.72	0.84	0.84	0.83	0.69	0.58	0.6	5743

5   Deployment and Architecture of Integrated System

The proposed model LSTM-LM is deployed in the flavour of following four different products for the convenience of end users by iShield.ai. (1) **Dost**[4]: This bot

---

[4] https://ishield.ai/dost.

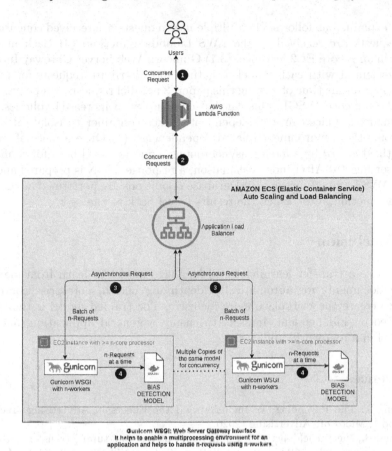

Fig. 2. Bias detection system architecture

can be configured to detect bias on enterprise communication platforms such as *Slack* and *Microsoft teams*. (2) **Chrome Plugin**[5]: This plugin can be configured with chrome browser for screening text contents for bias in any web applications. (3) **Content screener**[6]: This web application allows Checking for bias in created text contents before publishing them. (4) **Application Programming Interface (API)**[7]: This API can be integrated with enterprise platforms where contents are created and published.

The backend of the system is deployed on Amazon Web Services (AWS) and is built to scale for industry use. The current architecture can process 8 parallel requests. This architecture can be easily scaled to accommodate increased request volumes. The architecture of the backend system is shown in Fig. 2 and

---

[5] https://ishield.ai/chrome-plugin.
[6] https://ishield.ai/screener.
[7] https://ishield.ai/api.

can be explained as follows. (1) Multiple user requests are received concurrently. (2) Requests are received at the AWS Lambda Function. (3) Each model is placed in an n-core EC2 instance. (3.1) Gunicorn Web Server Gateway Interface is implemented with each model to gather and distribute requests for parallel processing. At the time of writing this paper, 8 parallel requests can be processed by each Gunicorn WSGI. This can be scaled up with increased volumes. (3.2) Each model is placed in an independent docker container to isolate it's function from other environment related dependencies. (4) Once a piece of content is identified *biased* by a model, asynchronous requests travel back for confidence based sorting. (5) After index calculation, a response JSON is prepared and sent to the AWS Lamba function. (6) Database operations are performed at the AWS Lamda Function, after which the results travel back to the user.

6   Conclusion

We propose a transfer learning based language model to learn from manually tagged documents for automatically identifying bias in enterprise content in order to create the workplace more inclusive. The trained model is thoroughly evaluated on independent datasets to ensure a general application, and it is deployed in a real-world application.

References

1. Gen, Z.: The Psychology of Inclusion and the Effects in Advertising. Technical report, Microsoft Advertising (2020)
2. Abul, B., NayakRichi: active learning for effectively fine-tuning transfer learning to downstream task. ACM Trans. Intell. Syst. Technol. (TIST) **12**(2) (2021). https://doi.org/10.1145/3446343
3. Agovino, T.: Toxic Workplace Cultures Are Costing Employers Billions. Technical report, Society for Human Resource Management (2019). https://www.talkworkculture.com/advice-info/toxic-workplace/
4. Attenberg, J., Provost, F.: Why label when you can search?: alternatives to active learning for applying human resources to build classification models under extreme class imbalance. In: Proceedings of the 16th ACM SIGKDD International Conference on Knowledge Discovery and Data Mining, pp. 423–432. ACM (2010)
5. Attenberg, J., Provost, F.: Inactive learning?: difficulties employing active learning in practice. ACM SIGKDD Explor. Newsl **12**(2), 36–41 (2011)
6. Bailinson, P., Decherd, W., Ellsworth, D., Guttman, M.: Understanding organizational barriers to a more inclusive workplace. Technical report, McKinsey Insights (2020)
7. Bashar, M., Nayak, R.: QutNocturnal@HASOC'19: CNN for hate speech and offensive content identification in Hindi language. In: CEUR Workshop Proceedings, vol. 2517 (2019)
8. Bashar, M.A., Nayak, R., Suzor, N., Weir, B.: Misogynistic tweet detection: modelling CNN with small datasets. In: Islam, R., et al. (eds.) AusDM 2018. CCIS, vol. 996, pp. 3–16. Springer, Singapore (2019). https://doi.org/10.1007/978-981-13-6661-1_1

9. Bashar, M.A., Nayak, R., Luong, K., Balasubramaniam, T.: Progressive domain adaptation for detecting hate speech on social media with small training set and its application to COVID-19 concerned posts. Social Netw. Anal. Mining **11**(1), 1–18 (2021). https://doi.org/10.1007/S13278-021-00780-W
10. Bashar, M.A., Nayak, R., Suzor, N.: Regularising LSTM classifier by transfer learning for detecting misogynistic tweets with small training set. Knowl. Inf. Syst., 1–26 (2020). https://doi.org/10.1007/s10115-020-01481-0
11. Bordia, S., Bowman, S.R.: Identifying and reducing gender bias in word-level language models. In: Proceedings of the 2019 Conference of the North, pp. 7–15. Association for Computational Linguistics, Stroudsburg (2019). https://doi.org/10.18653/v1/N19-3002, http://aclweb.org/anthology/N19-3002
12. Bradbury, J., Merity, S., Xiong, C., Socher, R.: Quasi-recurrent neural networks (2016). arXiv preprint arXiv:1611.01576
13. Lin, M.C.: Active learning with unbalanced classes & example-generated queries. In: AAAI Conference on Human Computation (2018)
14. Craven, M., Kumlien, J., et al: Constructing biological knowledge bases by extracting information from text sources. In: ISMB, vol. 1999, pp. 77–86 (1999)
15. Devlin, J., Chang, M.W., Lee, K., Toutanova, K.: Bert: pre-training of deep bidirectional transformers for language understanding (2018). arXiv preprint arXiv:1810.04805
16. Dinan, E., Fan, A., Williams, A., Urbanek, J., Kiela, D., Weston, J.: Queens are powerful too: mitigating gender bias in dialogue generation. In: Proceedings of the 2020 Conference on Empirical Methods in Natural Language Processing (EMNLP), pp. 8173–8188. Association for Computational Linguistics, Stroudsburg (2020). https://doi.org/10.18653/v1/2020.emnlp-main.656, https://www.aclweb.org/anthology/2020.emnlp-main.656
17. Dinan, E., Fan, A., Wu, L., Weston, J., Kiela, D., Williams, A.: Multi-Dimensional Gender Bias Classification. Technical report (2020)
18. Dixon-Fyle, S., Dolan, K., Hunt, V., Prince, S.: Diversity wins: how inclusion matters. Technical report, McKinsey & Company (2020). https://www.mckinsey.com/featured-insights/diversity-and-inclusion/diversity-wins-how-inclusion matters
19. Ehrenberg, H.R., Shin, J., Ratner, A.J., Fries, J.A., Ré, C.: Data programming with ddlite: putting humans in a different part of the loop. In: Proceedings of the Workshop on Human-In-the-Loop Data Analytics, p. 13. ACM (2016)
20. Garr, S.S., Jackson, C.: Diversity & inclusion technology: the rise of a transformative market. Technical report, Mercer, New York, United States (2019). https://www.mercer.com/our-thinking/career/diversity-and-inclusion-technology.html
21. Glorot, X., Bengio, Y.: Understanding the difficulty of training deep feedforward neural networks. In: Proceedings of the Thirteenth International Conference on Artificial Intelligence and Statistics, pp. 249–256 (2010)
22. Grosz, D., Conde-Cespedes, P.: Automatic Detection of Sexist Statements Commonly Used at the Workplace. Technical report (2020). https://hal.archives-ouvertes.fr/hal-02573576
23. Hearst, M.A., Dumais, S.T., Osuna, E., Platt, J., Scholkopf, B.: Support vector machines. IEEE Intell. Syst. Appl **13**(4), 18–28 (1998)
24. Herbert, F.: Is unconscious bias training still worthwhile? LSE Business Review (2021). https://blogs.lse.ac.uk/businessreview/
25. Hochreiter, S., Schmidhuber, J.: Long short-term memory. Neural Comput. **9**(8), 1735–1780 (1997)
26. Hoerl, A.E., Kennard, R.W.: Ridge regression: applications to nonorthogonal problems. Technometrics **12**(1), 69–82 (1970)

27. Hube, C., Fetahu, B.: Detecting biased statements in wikipedia. In: The Web Conference 2018 - Companion of the World Wide Web Conference, WWW 2018, pp. 1779–1786. Association for Computing Machinery Inc., New York (2018). https://doi.org/10.1145/3184558.3191640, http://dl.acm.org/citation.cfm?doid=3184558.3191640

28. Johnson, S.K., Hekman, D.R., Chan, E.T.: If There's Only One Woman in Your Candidate Pool, There's Statistically No Chance She'll Be Hired. Technical report, Harvard Business Review. https://hbr.org/2016/04/if-theres-only-one-woman-in-your-candidate-pool-theres-statistically-no-chance-shell-be-hired

29. Jozefowicz, R., Vinyals, O., Schuster, M., Shazeer, N., Wu, Y.: Exploring the limits of language modeling (2016). arXiv preprint arXiv:1602.02410

30. Kaneko, M., Bollegala, D.: Gender-preserving debiasing for pre-trained word embeddings. In: Proceedings of the 57th Annual Meeting of the Association for Computational Linguistics, pp. 1641–1650. Association for Computational Linguistics, Stroudsburg (2019). https://doi.org/10.18653/v1/P19-1160, https://www.aclweb.org/anthology/P19-1160

31. Lewis, D.D.: Naive (Bayes) at forty: the independence assumption in information retrieval. In: Nédellec, C., Rouveirol, C. (eds.) ECML 1998. LNCS, vol. 1398, pp. 4–15. Springer, Heidelberg (1998). https://doi.org/10.1007/BFb0026666

32. Liaw, A., Wiener, M., et al.: Classification and regression by randomForest. R news **2**(3), 18–22 (2002)

33. Manzini, T., Chong, L.Y., Black, A.W., Tsvetkov, Y.: Black is to criminal as caucasian is to police: detecting and removing multiclass bias in word embeddings. In: Proceedings of the 2019 Conference of the North, vol. 1, pp. 615–621. Association for Computational Linguistics, Stroudsburg (2019). https://doi.org/10.18653/v1/N19-1062, http://aclweb.org/anthology/N19-1062

34. Maudslay, R.H., Gonen, H., Cotterell, R., Teufel, S.: It's all in the name: mitigating gender bias with name-based counterfactual data substitution. In: EMNLP-IJCNLP 2019–2019 Conference on Empirical Methods in Natural Language Processing and 9th International Joint Conference on Natural Language Processing, Proceedings of the Conference, pp. 5267–5275 (2019). http://arxiv.org/abs/1909.00871

35. Merity, S., Xiong, C., Bradbury, J., Socher, R.: Pointer sentinel mixture models (2016). arXiv preprint arXiv:1609.07843

36. Mikolov, T., Grave, E., Bojanowski, P., Puhrsch, C., Joulin, A.: Advances in pre-training distributed word representations. In: Proceedings of the International Conference on Language Resources and Evaluation (LREC 2018) (2018)

37. Mikolov, T., Sutskever, I., Chen, K., Corrado, G.S., Dean, J.: Distributed representations of words and phrases and their compositionality. In: Advances in Neural Information Processing Systems, pp. 3111–3119 (2013)

38. Mikolov, T., Karafiát, M., Burget, L., Černockỳ, J., Khudanpur, S.: Recurrent neural network based language model. In: Eleventh Annual Conference of the International Speech Communication Association (2010)

39. Mitchell, M., et al.: Diversity and inclusion metrics in subset selection. In: Proceedings of the AAAI/ACM Conference on AI, Ethics, and Society 7 (2020). https://doi.org/10.1145/3375627

40. Patterson, G., Van Horn, G., Belongie, S.J., Perona, P., Hays, J.: Tropel: crowdsourcing detectors with minimal training. In: HCOMP, pp. 150–159 (2015)

41. Pryzant, R., Martinez, R.D., Dass, N., Kurohashi, S., Jurafsky, D., Yang, D.: Automatically neutralizing subjective bias in text. In: Proceedings of the AAAI Conference on Artificial Intelligence, vol. 34, pp. 480–489 (2020). https://doi.org/10.1609/aaai.v34i01.5385, https://github.com/rpryzant/neutralizing-bias

42. Recasens, M., Danescu-Niculescu-Mizil, C., Jurafsky, D.: Linguistic models for analyzing and detecting biased language. In: Proceedings of the 51st Annual Meeting of the Association for Computational Linguistics, pp. 1650–1659. Association for Computational Linguistics, Sofia (2013). http://en.wikipedia.org/wiki/

43. Safavian, S.R., Landgrebe, D.: A survey of decision tree classifier methodology. IEEE Trans. Syst. Man Cybern. **21**(3), 660–674 (1991). https://doi.org/10.1109/21.97458

44. Wang, T., Zhao, J., Yatskar, M., Chang, K.W., Ordonez, V.: Balanced Datasets are Not Enough: Estimating and Mitigating Gender Bias in Deep Image Representations. Technical report (2019)

45. Weinberger, K.Q., Saul, L.K.: Distance metric learning for large margin nearest neighbor classification. J. Mach. Learn. Res. **10**(Feb), 207–244 (2009)

46. Zalis, S.: Inclusive ads are affecting consumer behavior, according to new research. Technical report, Think with Google (2019). https://www.thinkwithgoogle.com/future-of-marketing/management-and-culture/diversity-and-inclusion/thought-leadership-marketing-diversity-inclusion/

47. Zmigrod, R., Mielke, S.J., Wallach, H., Cotterell, R.: Counterfactual data augmentation for mitigating gender stereotypes in languages with rich morphology. In: ACL 2019–57th Annual Meeting of the Association for Computational Linguistics, Proceedings of the Conference, pp. 1651–1661 (2019). http://arxiv.org/abs/1906.04571

Exploring Fusion Strategies in Deep Learning Models for Multi-Modal Classification

Duoyi Zhang[✉], Richi Nayak, and Md Abul Bashar

Queensland University of Technology, Brisbane, Australia
duoyi.zhang@hdr.qut.edu.au, {r.nayak,m1.bashar}@qut.edu.au

Abstract. When effectively used in deep learning models for classification, multi-modal data can provide rich and complementary information and can represent complex situations. An essential step in multi-modal classification is data fusion which aims to combine features from multiple modalities into a single joint representation. This study investigates how fusion mechanisms influence multi-modal data classification. We conduct experiments on four social media datasets and evaluate multi-modal models with several classification criteria. The results show that the quality of data and class distribution significantly influence the performance of the fusion strategies.

Keywords: Deep learning · Multi-modal data · Multi-modal fusion · Classification · Data mining

1 Introduction

With the popularity of heterogeneous networks such as the Internet of things, social networks and different digital publications, obtaining different types of data representing an event/entity has become facile. Data that contains different types/views representing the same subject is referred as *multi-modal* data, while data with a single type/view (e.g. images or text alone) is referred as *uni-modal data*. Figure 1 shows two examples of multi-modal data with textual and visual modalities.

Unlike uni-modal data, multi-modal data has richer and complementary information, and therefore it can represent complex situations in which different types of data are combined to show the full meaning the author likes to express. For example, the image in Fig. 1(a) shows the cover of a DVD, and the text below further expresses the author's feeling toward it. By image alone, the sentiment and context cannot be understood. Similarly, the text alone does not represent the situation effectively. The same can be observed in the second scenario. If not accompanied by the text, the image is simply showing a cat in Fig. 1(b). In multi-modal data, each modality merely provides part of the information about the topic. By combining both modalities, a machine learning model can understand the context and other embedded information.

© Springer Nature Singapore Pte Ltd. 2021
Y. Xu et al. (Eds.): AusDM 2021, CCIS 1504, pp. 102–117, 2021.
https://doi.org/10.1007/978-981-16-8531-6_8

I couldn't be more ecstatic today. I have been waiting 8 years
for 101 Dalmatians to release from the Disney Vault!

Watching all the flies go by! Which one shall I attack first?

(a)   (b)

Fig. 1. Examples of multi-modal data showing images and text data

Several deep learning models have been introduced for the task of multi-modal data classification leveraging the expressiveness of multi-modal data [1, 3,9,21]. A fundamental requirement for the success of these models is how to combine different modalities in their processing. Vanilla fusion, based on simple operations such as addition, pooling or concatenation of features of each modality, is one of the most commonly used methods [2,21,22]. However, this type of fusion fails to explore the relationships between modalities as they cannot control the importance of each modality in processing. For example, for the task of sentiment mining of the image and text in Fig. 1(a), the image based features are less informative than text features. The text features would deem more contributing to sentiment classification than the image based features in this data. Vanilla fusion mechanisms cannot accomplish this due to the lack of considerations about the interactions between multiple modalities. To overcome this, attention-based methods are proposed that achieves this by dynamically focusing on the informative modality. However, adding more interactions in the model leads to a more complex architecture, which results in various training problems such as overfitting, loss function optimization, etc. [20].

In this paper, we present a comprehensive study understanding the effect of fusion strategies in multi-modal classification. Unlike previous works that explore a specific fusion strategy with a targeted task [2,21,22], we present a general analysis of vanilla and attention-based fusion strategies for four common social media classification tasks including sentiment analysis, hate speech detection, disaster-related information detection, and fake news detection. Experiments on these four tasks reflect the capacity of multi-modal models using different fusion strategies, and suggest that the choice of fusion strategy should be determined by the characteristics of dataset. Additionally, we present this analysis with rigorous evaluation criteria, instead of accuracy only that is commonly used in prior research. Using complementary measures, the results can reflect the comprehensive distinguishability of classification models.

More specifically, the contributions of this paper are as follows.

1. We investigate the performance of multi-modal deep learning models in four different social media classification tasks using thorough evaluation criteria.

2. We experiment with vanilla and attention-based mechanisms to combine multiple modality features in deep learning models. The results show that the quality of data and the class distribution influence the performance of fusion mechanisms.

2   Related Work

Our focus in this paper is on multi-modal data classification with textual and visual modalities. Due to the individual success in natural language processing and computer vision, deep learning models have been found useful for the task of multi-modal classification [23]. In this section, we discuss recent works on deep learning-based multi-modal classification.

Researchers [1,12] have identified that, for some cases, only a single modality (i.e., image or text) is informative. That is, an image is not informative but the text is, and vice versa. In these cases, a uni-modal based architecture becomes biased if the less informative data is used in classification. Interestingly, a multi-modal architecture can learn from both modalities and improve classification accuracy due to embedding more expressive data features, in comparison to utilising the individual data modality in the models [1,3,9,21].

A fundamental requirement for multi-modal classification is how to fuse data from different modalities in a deep learning model to capture the dependencies of multiple modalities. Vanilla fusion strategy, which is one of the most popular fusion mechanisms, applies simple operations such as concatenation, addition and pooling to fuse features extracted from different modalities [15,17]. Concatenation directly links features in the next layer of a deep learning model whereas addition adds feature vectors together [15]. Bilinear pooling, i.e., calculating the outer-product between extracted features, is found to significantly improve f1-score for classification [17].

Although the vanilla fusion strategies show improved classification results in comparison to single modality models, they have two limitations. Firstly, studies show that the vanilla multi-modal models only have a marginally superior performance or sometimes even worse than the best uni-modal model [7,8]. This is probably because the vanilla mechanism can inevitably integrate noise from individual modality [7], which results in inaccurate predictions. Secondly, these methods treat all modalities equally in the model without leveraging the added contribution of a specific modality. For some tasks, however, one modality might contain more information than others [1,12]. To better distinguish between classes, the classifier is expected to learn more from one modality than the other.

An attention-based mechanism can achieve this by dynamically adjusting feature weights [23]. Prior research [10] has shown that the attention-based method has slightly better accuracy than the vanilla method for a large-scale data. Various studies have applied attention mechanisms to enhance the accuracy performance in a classification task [1,22]. Although an attention-based mechanism is shown to yield better classification performance, it inevitably increases the complexity of the multi-modal model including its architecture. Consequently, these

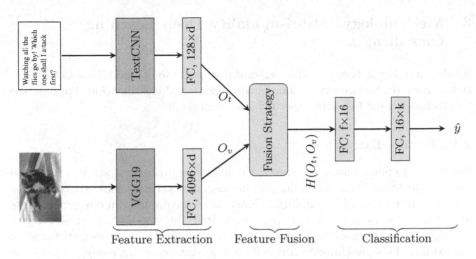

Fig. 2. Architecture of the multi-modal data classification. d and f are the dimensions of the extracted features for image/text modalitis and the fused features respectively. k is the number of classes. FC is a fully connected network. For example, it projects the text features extracted from the TextCNN network (128 dimensions) to a lower-dimensional space (d dimensions).

models require a larger training dataset and a rigorous training regime. Prior research has shown that obtaining a large-scale annotated dataset is difficult and infeasible sometimes [13].

With the vast popularity of deep learning models and multi-modality datasets, it is significant to understand the performance of these models under various fusion strategies. This will help deep learning research field to innovate modelling architectures and strategies. In literature, the majority of works focus on designing and comparing fusion mechanisms for a specific task without understanding the effects [1,22]. As a result, a multi-modal method is hard to transfer to other tasks due to the lack of generalization. Moreover, while comparing different multi-modal classification models, accuracy and f1-score are used as the common evaluation criteria. However, most of the existing multi-modal datasets are imbalanced. It is important to study these methods with additional measurement metrics for rigorous evaluation because some metrics can be biased (e.g. accuracy can be biased to the majority class).

In this paper, we assess several fusion strategies on four multi-modal datasets representing different social media tasks with six varied classification criteria. To our best of knowledge, this will be the first empirical study presenting a rigours comparison of deep learning models with different fusion strategies.

3 Methodology: Multi-modality Deep Learning Classification

As shown in Fig. 2, there are three essential steps for multi-modal data classification, which are feature extraction, feature fusion and classification. We describe the techniques used in this paper for empirical study.

3.1 Feature Extraction

For the text component of a dataset, the following processing was conducted to prepare the text data for modelling. Punctuation was removed and shorten forms of words were replaced by full forms. Since all of the datasets in our experiments are Twitter datasets, for the pre-processed data, the state-of-the-art GloVe [16] pre-trained on a large-scale Twitter data with 200 dimensions was used for word embedding. This pre-trained word-embedding model contains many social media tokens, such as "RT" for re-tweet, <url> for URLs, and <hashtag> for hashtags. Since the size of the labelled data available for training the social media tasks is not large enough, we do not fine-tune the weights for word embeddings to prevent overfitting. Furthermore, we applied TextCNN [11] for textual feature extraction. By using various window sizes, kernels for TextCNN can learn different granularity features about the text. After that we use a fully connected layer to project the TextCNN features into a lower dimensional space. i.e.:

$$O_t = ReLu(W_t F_{textcnn}) \tag{1}$$

where $F_{textcnn}$ is the feature obtained from TextCNN, $ReLu$ is the activation function, W_t is the weight matrix for the fully connected layer, O_t is the output features, $O_t \in \mathbf{R}^d$, and d is the dimension of extracted textual features. We treat d as a hyperparameter and tune it accordingly for each dataset.

For the image component of a dataset, the following processing was conducted to prepare the image data for modelling. Due to the success in accomplishing computer vision tasks, the state-of-the-art VGG19 [18] network, pre-trained on the large imagenet dataset [6], is applied for image feature extraction. We replace the last four fully connected layers in the VGG19 network with a single fully connected layer. The aim of using additional fully connected layer is to project the features obtained from VGG19 to a lower space with the equivalent number of dimension as the extracted text features, which make both modalities have equal contributions in the modelling process. This layer is added to project features into a lower dimension. Moreover, we do not fine-tune the VGG19 model to prevent overfitting. For each image v_i:

$$O_v = ReLu(W_v F_{vgg19}) \tag{2}$$

where W_v is the leanable parameters for the fully connected layer, F_{vgg19} is the feature obtained from VGG19, O_v is the output features, $O_v \in \mathbf{R}^d$.

3.2 Fusion Strategy

Once the features are ready to be processed, the next step is combining the features from both modalities. Two vanilla methods, as shown in Figs. 3(a) and 3(b), and two attention-based methods, as shown in Figs. 3(c) and 3(d) are used in combining features from different modalities.

Concatenation. Concatenation means features from both modalities are connected and directly fed into the next layer of the network. The joint representation of textual and visual features can be presented as,

$$H(O_t, O_v) = ReLu(W[O_t|O_v] + b) \tag{3}$$

where | means concatenation, W is the weight matrix, b is the bias, ReLu is the activation function, $H(O_t, O_v) \in \mathbf{R}^f$ and f is dimension of the joint features.

Addition. In this fusion strategy, image and text features are combined by element-wise addition. The joint representation of textual and visual features can be presented as,

$$H(O_t, O_v) = ReLu(W(O_t + O_v) + b) \tag{4}$$

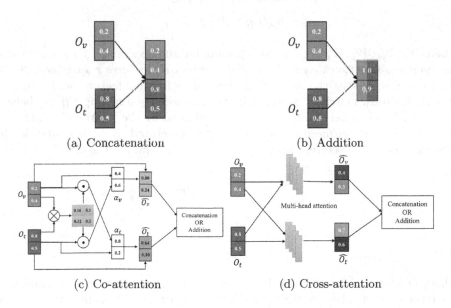

(a) Concatenation   (b) Addition

(c) Co-attention   (d) Cross-attention

Fig. 3. Architectures of vanilla and attention-based fusion strategies. It also shows a simple example of how vectors from each modality are combined and a fused vector is obtained.

Co-Attention. We use a parallel co-attention mechanism [10,14] which is originally proposed for the task of visual question answering. Different from classification, this task focuses on answering questions from the provided visual information. In other words, it aims to align each token in the text with a location in the image. Therefore, the first step in the original algorithm is calculating the similarity matrix between image and text features at all image locations. However, in classification, text-image pair may not have this location-based correlation. Instead of calculating the similarity, we firstly captured all the interactions between textual and visual vectors. The following steps remain the same.

Initially, we calculate the outer-product of image and text features to capture all the interactions between modalities, i.e.: $P = O_t^T \otimes O_v$. Therefore, row_i in matrix P represents the interaction between textual feature i and all of the visual features, wheares $column_j$ in matrix P represents the interaction between visual feature j and all of the visual features.

Equation 5 is applied to combine the features and calculate the attention weights vectors accordingly. Specifically, it combines uni-modal features with the interaction enhanced features. Then the attention weight vectors α_v, α_t are calculated using the combined features H_v and H_t.

$$
\begin{aligned}
H_v &= tanh(W_v O_v + (W_t O_t)P) \\
\alpha_v &= Softmax(W_v' H_v) \\
H_t &= tanh(W_t O_t + (W_v O_v)P^T) \\
\alpha_t &= Softmax(W_t' H_t)
\end{aligned}
\tag{5}
$$

where W_v, W_v', W_t, W_t' are learnable parameter matrices, and α_v, α_t are vectors representing attention weights for visual and textual features respectively. Since the column represents the interaction between each visual features and all of the textual features, matrix P needs to be transposed. By applying Eq. (6) below, the textual and visual feature vectors are weighted by the attention weights. In general, co-attention mechanism uses weights generated by both modalities to dynamically adjust the representation of the modal-specific features.

$$
\begin{aligned}
\hat{O}_v &= \sum_{n=1}^{d} \alpha_v^n * O_v^n \\
\hat{O}_t &= \sum_{n=1}^{d} \alpha_t^n * O_t^n
\end{aligned}
\tag{6}
$$

where $\hat{O}_v$ and $\hat{O}_t$ are attended image and text features, respectively, and d is the dimension of extracted features. Finally, the attended features can be either concatenated by Eq. (3) or added by Equation (4) to get a joint representation.

Cross-Attention. Cross-attention is another dynamic weighted sum approach, which allows one modality to influence the other ones. Specifically, the weights for image features is calculated using text features, and vice versa. We employ a multi-head attention layer [19] to generate the weighted sum representing the fused text and image modalities. Instead of generate a single weight vector, multi-head can generate multiple attention vectors in parallel. After that, we concatenate these results, i.e.:

$$\alpha_v = [Head_v^1|...|Head_v^h]W_v$$

$$Head_v^i = Softmax(\frac{O_v * O_t^T}{\sqrt{d}})$$

$$\alpha_t = [Head_t^1|...|Head_t^h]W_t \qquad (7)$$

$$Head_t^i = Softmax(\frac{O_t * O_v^T}{\sqrt{d}})$$

where α_v and α_t represent attention weight vectors for image and text respectively, W_v, W_t are weight matrices for images and text data respectively, and h is a hyperparameter representing the number of heads.

Similar to co-attention. the image and text features are then attended to or weighted by Eq. (6). Finally, the attended features are concatenated by Eq. (3) or added by Eq. (4) for the joint representation. There are two differences between cross-attention and co-attention. Firstly, cross-attention only applies the interaction between inter model features to calculate the attention weights, whereas co-attention combines modal-specific features with interactive features. In other words, co-attention involves inter-modal information as well as modal-specific information. Secondly, unlike co-attention which only generate a single attention weight, cross-attention can generate it multiple times.

3.3 Classification

As shown in Fig. 2, the final step is classification, which is based on a multi-layer perceptron as shown in Eq. (8). We use softmax function for multi-class classification, and sigmoid function for binary classification.

$$\hat{y} = Softmax(WH(O_t, O_v) + b)$$

$$\hat{y} = Sigmoid(WH(O_t, O_v) + b) \qquad (8)$$

where W is the weight matrix and $H(O_t, O_v)$ is the joint features.

3.4 Complexity Analysis of a Fusion Strategy

It is hard to have a precise complexity analysis for the entire multi-modal model [10]. We roughly analyse the complexity of the fusion mechanism. Let k be the number of classes and d be the number of dimensions of textual/visual features. The complexity of concatenation and addition will be $O(2dk)$ and $O(dk)$, respectively. Both the co-attention and cross-attention involve calculating the dot product between features resulting in higher complexity. For co-attention, we directly calculate the dot-product between two features, resulting in the complexity of $O(d^2k)$. For cross-attention, however, due to the use of multiple parallel headings, the complexity is $O(hd^2k)$, which is slightly higher. In conclusion, a vanilla algorithm takes linear time to run, while an attention-based algorithm costs quadratic time.

4   Experiments

This paper aims to investigate the influence of different fusion strategies towards different classification tasks with social media data. We first introduce the datasets and classification criteria to evaluate a model.

Table 1. Details of four social media datasets

Dataset	Task	Size (train/test)	Number of classes	Ratio
MVSA-single	Sentiment detection	4511 (4059/452)	3	59% Positive, 30% Negative,10% Neutral
Hate speech	Hate speech detection	5000 (3998/1002)	4	37.0%Hateful,7.3%Reclaimed,2.3%Counterspeech,53.4%None
CrisisMMD	Crisis event detection	12708 (11174/1534)	2	38.7% Non informative, 61.3%Informative
MediaEval2016	Fake news detection	12,291 (11,519/772)	2	44.2% Real, 55.8%Fake

4.1   Datasets

Since most recent multi-modal data classification tasks targeted at social media platforms [2,21], we evaluate the classification performance on four twitter datasets that contain image-text pairs representing a social media task. Details of the four datasets used in this research are summarized in Table 1.

Sentiment Analysis (MVSA): This dataset [21] contains tweets with images and text for sentiment analysis. Given a tweet, the objective is to identify whether it is of positive, negative, or neutral sentiment. The original dataset provides annotations for the image and text independently and labels for tweets representing both modalities need to be generated. In this paper, we followed steps in [21] to annotate tweets with unmatched text and image labels.

Hate Speech Detection: This dataset [3] contains 5000 image-text pairs. The text comes from two sources: from the tweet and from the image. The dataset is originally split into training, validation, and test sets, and we use the same splitting.

Disaster Detection (CrisisMMD): This dataset [2] contains tweets with images and text for identification of natural disaster-related tweets. The goal is to identify if a tweet, which is an image-text pair, can provide useful information, such as damage to buildings or injured people. In other words, it aims to classify a tweet as informative or non-informative. Aside from the annotation for the entire tweet, image and text content has also been annotated separately. Consequently, images and text content might have different labels. In this work, we considered tweets with unmatched labels as noise, and they have been removed.

Fake News Detection (Mediaeval2016): The dataset [4] is about fake news detection for visual-text pairs. The visual content contains videos and images. Since we only focused on the image and text data, we removed instances containing videos. The dataset is originally split into non-overlapping training and test sets.

4.2 Baseline Models

We use two uni-modal models, based on the text-only modality (i.e., the TextCNN language model) and based on the image only modality (i.e., the VGG19 visual model), as baseline models. We also included the best performance reported in the literature on these datasets [1,3,9,21]. These publications may not provide results in all the criteria as used in this paper. Moreover, they didn't perform 5-fold validation as our experiments on each dataset.

4.3 Experimental Set-Up

Except for the MVSA dataset, all the datasets have provided the test set. To compare with previous works, we randomly generated 10% data as the test set for MVSA, and then hold out all the test sets. After that, to assess the generalizability, we implement 5-fold validation for the rest of the data. We applied the Stratified k-fold package from scikit-learn for splitting training and validation set for the 5-fold validation. We resized all images to 224×224 and padded texts to the maximum length. The window sizes for TextCNN are 2,3, and 4. As for classification, we used two fully connected layers with hidden sizes of f and 16, respectively. Besides, we applied cross-entropy and binary cross-entropy for multi-class classification and binary classification respectively. Furthermore, to reduce overfitting, we also include L_2 normalization with a scale factor λ which is a hyperparameter tuned in {1e-3,1e-4}. For cross-attention, the number of heads h is tuned between 1 and 8. We also tune the following hyperparameters: learning rate in {0.01, 0.05, 1e-3, 5e-3, 1e-4, 5e-4,1e-5}, batch size in {8, 16, 32, 64, 100, 128}, dimension of extracted features d in {8, 16, 32, 64, 128}, patience of scheduler in {5, 10, 15, 20, 30}, early stopping patience in {10, 20}. We conducted the experiments on HPC platform and implemented all models by Python 3.7.4 and Pytorch 1.3.1.

4.4   Evaluation Criteria

Since all the datasets in our experiments are imbalanced, the overall accuracy might be biased towards the majority class. For this reason, we also evaluate models on other five evaluation criteria. We considered six evaluation criteria, namely accuracy, f1-score, precision, recall, cohen-kappa (CK), and AUC [5]. Since macro average can alleviate the influence of the majority class, we consider macro averaged precision, recall, f1-score, and AUC score. In addition, CK evaluates the number of predictions a model makes that cannot be explained by the random guessing [5]. In other words, it can reflect the improvement from predicting everything as the majority class. Additionally, due to the use of 5-fold validation, we demonstrate the mean and standard deviation of five runs. Since the mean value can reflect the average performance of the model, the comparison is based on the mean value.

5   Results and discussion

In this section, we report the DNN model performances, summarize the results and discuss our findings.

5.1   Sentiment Analysis

Table 2 lists the result of the MVSA dataset for sensitivity analysis. Results show that all the multi-modal models outperform their uni-model counterparts in terms of all six criteria. CK and precision, in particular, have a significant improvement. It seems that the combination of modalities can provide more information for the classification. Thus, the classification model can better distinguish between sentiments.

Table 2. Results for MVSA dataset: sentiment analysis

		Accuracy	CK	Precision (macro)	Recall (macro)	F1 (macro)	AUC (macro)
Previous work	MultiSentiNet[21]	0.70	-	-	-	-	-
Baselines (single-modality)	VGG19	0.62±0.02	0.29±0.02	0.53±0.02	0.52±0.01	0.52±0.01	0.64±0.01
	TextCNN	0.64±0.01	0.34±0.02	0.55±0.01	0.52±0.01	0.52±0.01	0.65±0.01
Fusion strategy	Concatenation	0.69±0.01	0.39±0.02	0.60±0.03	0.55±0.02	0.56±0.02	0.67±0.02
	Addition	0.70±0.02	0.41±0.02	0.61±0.02	0.54±0.02	0.56±0.01	0.67±0.01
	Co-attention concatenation	0.68±0.02	0.39±0.02	0.58±0.04	0.55±0.02	0.55±0.02	0.68±0.01
	Co-attention addition	0.69±0.01	0.39±0.03	0.61±0.03	0.54±0.02	0.55±0.01	0.67±0.01
	Cross-attention concatenation	0.70±0.01	0.42±0.02	0.61±0.02	**0.56±0.01**	0.57±0.01	0.68±0.01
	Cross-attention addition	**0.71±0.01**	**0.44±0.02**	**0.65±0.03**	**0.56±0.01**	**0.58±0.01**	**0.69±0.01**

In terms of fusion strategies, vanilla methods and co-attention have relatively close performance. As illustrated in Table 5, the macro-averaged recall is low. Since this dataset has three imbalanced classes, this result indicates that the classifier tends to learn from the majority classes, which are positive and negative, and ignore the neutral class. The previous work [21] on this dataset applied the concatenation model, which has almost the same accuracy as our vanilla method. In addition, the cross-attention with addition model outperform others. In particular, it achieves the highest accuracy and precision rate.

5.2   Hate Speech Detection

Table 3. Results for Hate speech dataset

		Accuracy	CK	Precision (macro)	Recall (macro)	F1 (macro)	AUC (macro)
Previous work	BERT-like fusion[3] (Weighted average)	0.79	-	0.76	0.79	0.77	-
Baselines (single-modality)	VGG19	0.53±0.01	0.17±0.01	0.37±0.02	0.35±0.00	0.36±0.01	0.57±0.00
	TextCNN	0.72±0.02	0.49±0.04	**0.58±0.12**	**0.45±0.03**	**0.47±0.04**	**0.66±0.02**
Fusion strategy	Concatenation	**0.73±0.01**	**0.51±0.03**	0.54±0.11	**0.45±0.02**	0.46±0.03	**0.66±0.01**
	Addition	0.72±0.01	0.48±0.01	0.41±0.07	0.40±0.01	0.39±0.02	0.64±0.01
	Co-attention concatenation	**0.73±0.01**	0.48±0.01	0.38±0.02	0.40±0.00	0.38±0.01	0.63±0.00
	Co-attention addition	**0.73±0.01**	0.49±0.01	0.37±0.00	0.40±0.00	0.38±0.00	0.63±0.00
	Cross-attention concatenation	**0.73±0.02**	0.50±0.03	0.46±0.07	0.42±0.02	0.41±0.03	0.65±0.02
	Cross-attention addition	0.72±0.01	0.48±0.03	0.53±0.05	0.41±0.01	0.41±0.01	0.64±0.01

Results of the hate speech dataset are shown in Table 3. The performance of the language model is marginally superior to multi-modal models for all criteria except accuracy and CK. In this dataset, the inclusion of visual information with the text data brings the adverse effect and a multi-modal results in inferior performance than the language based uni-model. As described in the previous section, images in this dataset embed with text, which hinders the model to learn informative visual features, such as features of an object or a scenario. This influence can also be reflected by the significantly worse performed image-only model. Hence, the inclusion of noisy visual features does not improve the classification.

A comparison between fusion methods indicate that processing features with more complex fusion methods results in inferior performances. The attention-based models have the worst macro-averaged precision and recall, however, they obtain the best accuracy performance. It may be due to the following two reasons. Firstly, the small number of training set might result in overfitting. There are only 3998 instances for training, from which we further split 20% for testing. Eventually, only about 3199 instances are used for model construction. Comparing with concatenation, attention-based models require more parameters. Consequently, the attention-based model tends to overfit the majority class, which can be proved by the lower macro-averaged f1-score. The higher performance in accuracy measure also reinforces this. Secondly, except for the overfitting problem, the interaction between modalities might feed misleading information for classification. Ideally, attention-based methods can reduce noise through the interaction between modalities. In this situation, however, image is too noisy to guide the fusion. Previous work [3] on this dataset shows weighted averaged results instead of macro-averaged. As suggested by accuracy, this model achieves significantly better performance than ours. They applied a pre-trained transformer to fuse the multiple modalities. Specifically, the visual features are projected into the text embedding space through BERT, which provides contextual information. Our vanilla and attention-based mechanisms, on the other hand, lack the assistant of contextual information.

5.3   Crisis Event Detection

Table 4. Results for crisismmd dataset: crisis event detection

		Accuracy	CK	Precision (macro)	Recall (macro)	F1 (macro)	AUC (macro)
Previous work	SSE-Cross-BERT-DenseNet[1]	0.89	-	-	-	0.88	-
Baselines (single-modality)	VGG19	0.83±0.00	0.61±0.01	0.81±0.01	0.81±0.01	0.81±0.01	0.81±0.01
	TextCNN	0.84±0.00	0.62±0.01	0.82±0.01	0.80±0.01	0.81±0.01	0.81±0.01
Fusion strategy	Concatenation	**0.88±0.00**	0.71±0.01	**0.87±0.01**	0.84±0.01	0.85±0.01	0.85±0.01
	Addition	0.87±0.00	0.71±0.00	**0.87±0.00**	0.84±0.00	0.85±0.00	0.84±0.00
	Co-attention concatenation	**0.88±0.00**	**0.72±0.01**	**0.87±0.01**	**0.86±0.01**	**0.86±0.00**	0.86±0.01
	Co-attention addition	**0.88±0.01**	**0.72±0.01**	**0.87±0.01**	**0.86±0.01**	**0.86±0.01**	**0.87±0.01**
	Cross-attention concatenation	0.87±0.00	0.70±0.01	0.86±0.01	0.84±0.00	0.85±0.00	0.84±0.00
	Cross-attention addition	0.87±0.00	0.71±0.01	0.86±0.00	0.85±0.01	0.85±0.00	0.85±0.01

Results for the CrisisMMD dataset are shown in Table 4. Compared with the uni-modal models, fusion-based methods can improve classification performance. The good performance of both uni-modal models indicates the high quality and less noisy data. The fusion-based methods can take advantage of the complementary nature of multi-modal data, and therefore all multi-modal models can improve the classification performance.

All multi-modal models achieve high performance in terms of all six criteria. Aside from the high quality data, another reason is this task is binary classification with almost balanced data, which is easier than previous two tasks. Previous work [1] achieves the highest accuracy and f1-score. Aside from the attention-based fusion method, they also included additional knowledge-graph based restrictions to enhance the embedding layer. In addition, the co-attention method achieves the best overall performance amongst all fusion-based methods. The higher recall, in particular, indicates that this method can find crisis-related tweets more frequently. Furthermore, the performance of cross-attention is near to the vanilla methods' performance. Comparing with cross-attention, co-attention captures both modal-specific and inter-modal features.

5.4   Fake News Detection

Table 5. Results for Mediaeval2016 dataset: fake news detection

		Accuracy	CK	Precision (macro)	Recall (macro)	F1 (macro)	AUC (macro)
Previous work	MVAE[9]	0.75	-	-	-	-	-
Baselines (single-modality)	VGG19	0.57±0.02	0.18±0.04	0.62±0.04	0.59±0.02	0.56±0.03	0.59±0.02
	TextCNN	0.58±0.03	0.13±0.05	0.57±0.03	0.56±0.03	0.56±0.03	0.56±0.03
Fusion strategy	Concatenation	0.69±0.01	0.38±0.02	0.70±0.01	0.70±0.01	0.69±0.01	0.70±0.01
	Addition	0.67±0.01	0.37±0.02	0.73±0.00	0.70±0.01	0.67±0.01	0.69±0.01
	Co-attention concatenation	**0.74±0.02**	**0.47±0.04**	0.74±0.02	**0.73±0.02**	**0.73±0.02**	**0.73±0.02**
	Co-attention addition	0.73±0.01	0.43±0.04	**0.78±0.01**	0.70±0.02	0.70±0.02	0.70±0.02
	Cross-attention concatenation	0.70±0.00	0.37±0.01	0.71±0.01	0.68±0.00	0.68±0.01	0.68±0.00
	Cross-attention addition	0.68±0.02	0.37±0.04	0.69±0.01	0.69±0.02	0.68±0.02	0.69±0.02

Results for the Mediaeval dataset are illustrated in Table 5. All the multi-modal models outperform baselines indicating that multi-modal models are more informative and leveraging inter-dependencies between two modalities. The performances of both uni-modal models on this dataset are undesirable. After reviewing the raw tweets, we recognize that many hashtags and URLs which are removed while pre-processing. As a result, the text might be too short to extract information. For the image, due to the replicated use of images, there are only almost 400 unique images for this dataset, which result in poor performance.

The better performance of vanilla models indicates that the two uni-modalities can complement each other and can produce better results when combined. In addition, as mentioned in [9], fake news has the characteristic of inconsistent text and images. By exploiting the interactions between modalities, the attention-based method can successfully capture this characteristic. By considering both inter-modal and modal-specific information, co-attention based models can better capture the above two characteristics.

The previous [9] deep learning model using concatenation fusion strategy has marginally improved accuracy in comparison to co-attention concatenation method. This research is based on variation auto-encoder [9] that has additional restrictions of reconstructing from the extracted features.

5.5   Final Remarks

Here, we summarise the key observations resulted from our experiments with vanilla and attention-based mechanisms for four social media datasets. In general, both of them can improve the classification performance for most tasks, which is consistent with previous works [2,21,22]. One significant result is that if one of the modalities is extremely noisy and there are a limited number of instances in the dataset, the performance of multi-modal models becomes close to or worse than the best uni-modal model. The fact that attention-based methods can also cause this under-performance phenomenon contradicts to the results of previous works [7,8] which shows that this phenomenon only appears in vanilla methods. Therefore, it is also challenging to design the attention-based multi-modal model when one of the modalities is noisy.

Another significant result is that co-attention illustrates a dominant performance for binary classification with balanced data. This is because the co-attention mechanism can balance the contribution of the modalities and capture the cross-modal features. This finding consists with previous works [1,9]. However, for the two multi-class social media tasks with imbalanced data, the four fusion methods show close performance and they tend to ignore the rare classes. Previous work [3] shows that pre-trained BERT-like fusion mechanism can improve the classification performance for this scenario. Specifically, they treated image features as special word tokens, and then fed the image as well as other text tokens into a pre-trained BERT model. However, this fusion mechanism might not work for non-textual multi-modal data, such as RGB-D data. Therefore, a more generalized fusion mechanism that can identify the deeper semantic connections between modalities needs to be designed. Besides, in this scenario, the concatenation is slightly better than the others. Moreover, experiments suggest that the concatenation is in general better than addition for the four classification tasks. Comparing two attention-based methods, co-attention is better for binary classification while cross-attention is good at multi-class classification. Besides, the complexity of the fusion methods might influence the multi-modal data classification. For a task with limited instances, such as hate speech detection, vanilla fusion strategies perform better.

6   Conclusion

In this paper, we investigate the multi-modal classification task with image and text modalities. The experiment results in terms of six criteria can work as baselines for the future works. We found that, in general, combining modalities can improve classification performance. In addition, the experiment results suggest different use scenarios for the vanilla and attention-based mechanisms.

References

1. Abavisani, M., Wu, L., Hu, S., Tetreault, J., Jaimes, A.: Multimodal categorization of crisis events in social media. IEEE (2020)
2. Alam, F., Ofli, F., Imran, M.: Crisismmd: multimodal twitter datasets from natural disasters. In: Proceedings of the International AAAI Conference on Web and Social Media, vol. 12 (2018)
3. Austin, V.B., Hale, S.A.B.: Deciphering implicit hate: evaluating automated detection algorithms for multimodal hate (2021)
4. Boididou, C., et al.: Verifying multimedia use at mediaeval (2016)
5. Czodrowski, P.: Count on kappa. J. Comput.-Aided Molec. Des. 28, 1049–1055 (2014)
6. Deng, J., Dong, W., Socher, R., Li, L.J., Li, K., Fei-Fei, L.: Imagenet: a large-scale hierarchical image database. IEEE (2009)

7. Jin, P., Li, J., Mu, L., Zhou, J., Zhao, J.: Effective sentiment analysis for multimodal review data on the web. In: Qiu, M. (ed.) ICA3PP 2020. LNCS, vol. 12454, pp. 623–638. Springer, Cham (2020). https://doi.org/10.1007/978-3-030-60248-2_43

8. Jin, Z., Cao, J., Guo, H., Zhang, Y., Luo, J.: Multimodal fusion with recurrent neural networks for rumor detection on microblogs. ACM (2017)

9. Khattar, D., Goud, J.S., Gupta, M., Varma, V.: MVAE: multimodal variational autoencoder for fake news detection. ACM (2019)

10. Kiela, D., Grave, E., Joulin, A., Mikolov, T.: Efficient large-scale multi-modal classification. In: Proceedings of the AAAI Conference on Artificial Intelligence, vol. 32 (2018)

11. Kim, Y.: Convolutional neural networks for sentence classification. Association for Computational Linguistics (2014)

12. Kumari, K., Singh, J.P., Dwivedi, Y.K., Rana, N.P.: Towards cyberbullying-free social media in smart cities: a unified multi-modal approach. Soft Comput. **24** (2020)

13. Lin, D., Li, L., Cao, D., Lv, Y., Ke, X.: Multi-modality weakly labeled sentiment learning based on explicit emotion signal for Chinese microblog. Neurocomputing **272** (2018)

14. Lu, J., Yang, J., Batra, D., Parikh, D.: Hierarchical question-image co-attention for visual question answering, pp. 289–297. Curran Associates Inc. (2016)

15. Madichetty, S., Muthukumarasamy, S., Jayadev, P.: Multi-modal classification of twitter data during disasters for humanitarian response. J. Ambient Intell. Hum. Comput. **12**, 1022–10237 (2021)

16. Pennington, J., Socher, R., Manning, C.: Glove: global vectors for word representation. Association for Computational Linguistics (2014)

17. Pranesh, R.R., Shekhar, A., Kumar, A.: Exploring multimodal features and fusion strategies for analyzing disaster tweets (2021)

18. Simonyan, K., Zisserman, A.: Very deep convolutional networks for large-scale image recognition (2015)

19. Vaswani, A., et al.: Attention is all you need, vol. 30. Curran Associates, Inc. (2017)

20. Wang, W., Tran, D., Feiszli, M.: What makes training multi-modal classification networks hard? IEEE (2020)

21. Xu, N., Mao, W.: Multisentinet. ACM (2017)

22. Xue, J., Wang, Y., Tian, Y., Li, Y., Shi, L., Wei, L.: Detecting fake news by exploring the consistency of multimodal data. Inf. Process. Manag. **58**, 102610 (2021)

23. Zhang, C., Yang, Z., He, X., Deng, L.: Multimodal intelligence: representation learning, information fusion, and applications. IEEE J. Sel. Topics Signal Process. **14**, 478–493 (2020)

Application Track

Chameleon: A Python Workflow Toolkit for Feature Selection

Diviya Thilakeswaran[1], Simon McManis[2], and X. Rosalind Wang[1,3]($\boxtimes$) (iD)

[1] Western Sydney University, Parramatta, NSW, Australia
`rosalind.wang@westernsydney.edu.au`
[2] University of Technology, Sydney, Australia
[3] CSIRO Data61, Epping, Australia

Abstract. When considering classification problems in relation to high-dimensional data sets, such as biological data sets, the need for effective methods of dimensionality reduction by feature selection becomes apparent. Feature selection has been shown to significantly decrease computational cost and allow for classification models that are more easily interpretable. We present *Chameleon*, a Python-based toolkit that integrates all steps in a feature selection evaluation pipeline – from splitting data for cross-validation, to visualisation of classification results using various metrics. We implemented in *Chameleon* six existing feature selection methods, six common classification methods, and the classification results are evaluated using two different metrics. We also implemented an ensemble method which selects only common features from the different methods evaluated. Experimental results using four different data sets suggest that the common features method achieves improved or similar classification performance, compared to the individual feature selection algorithms, using smaller and thus more computationally efficient subsets of features.

Keywords: Feature selection · Biological data · Classification

1 Introduction

The exponential growth in the volumes of data being generated in the world today has given rise to many data sets with extremely high dimensionality, posing the challenge of how to effectively extract the most relevant and valuable information from them. The application of data mining and machine learning techniques in order to uncover meaningful information from a data set becomes more complex as its dimensionality grows. A great increase in the number of features may lead to a significant amount of noise in the data, as well as the over-fitting and poor performance of learning models, in what is known as the curse of dimensionality [13].

Y. Xu et al. (Eds.): AusDM 2021, CCIS 1504, pp. 121–135, 2021.
https://doi.org/10.1007/978-981-16-8531-6_9

Yu and Liu [26] describe features as belonging to one of four categories; completely irrelevant features, weakly relevant and redundant features, weakly relevant and non-redundant features and strongly relevant features. Relevant features may have significant correlation with the predictors, and as such should be included in model building in order to increase prediction accuracy. Redundant features may correlate significantly with each other, and thus should be eliminated in order to improve model performance.

In order to simplify and improve the accuracy of classification models that are built upon these high-dimensional data sets, methods that can account for such irrelevant and redundant features in the data must be considered. This accounting for noise in a data set can be achieved by selecting a subset of features that are determined to be most relevant to the classification model being constructed. This effectively reduces the dimensionality of the data in a technique known as feature selection.

The importance of feature selection as a method of dimensionality reduction can be seen through its application in bioinformatics and medicine. In order to provide a more accurate understanding of underlying processes in such fields it may be imperative to retain the interactions of the original features [15]. Methods of feature selection have been shown to be effective in various applications, including improving classification performance with medical data [24].

1.1   Feature Selection for Classification

One of the most common applications for feature selection methods includes its use in classification problems [8,11]. Real-world data sets often contain a high multitude of different features, many of which are not relevant to the classification problem that is attempting to be solved. Training classifiers on high dimensional data sets with large amounts of irrelevant and redundant features is costly and unwieldy. Through the selection of an optimal subset of features, the subsequent process of building a classifier on these features becomes more cost efficient, and the resulting model is more generalisable.

Feature selection techniques for the purposes of classification can be categorised into two main categories: filter model and wrapper model. Filter models determine feature subsets based upon the general and intrinsic characteristics of the data [7]. Such methods are known to computationally efficient and are separate from classifier learning algorithms, thus they are able to be applied to many classifiers with better generalisation. The process of selecting feature subsets using filter methods is independent of learning algorithms, thus the selected subset may not provide the most optimal performance for the target classification model.

Wrapper models utilise specific learning algorithm and select optimal subsets based on classifier's prediction error [6]. As such, wrapper models are able to drive better classification performance compared to filter methods, though this comes at significant computational cost. Unlike filter methods, wrapper methods are less generalisable to learning algorithms other than the one used to determine their evaluation criterion, and are more likely to over-fit.

1.2   Python Toolkit for Feature Selection and Classification

Currently, there are few existing open source Python libraries for feature selection. The Python *Scikit-Learn* library is an extremely popular machine learning library, however the feature selection module contained within the library implements only a few methods [22]. Currently, the largest feature selection library for Python to exist is *Scikit-Feature* developed by Arizona State University, which contains around 40 algorithms for feature selection [20]. However, *Scikit-Feature* is implemented in Python 2, which is no longer under development.

Despite the existence of a few Python tools which contain various feature selection algorithms, there does not evidently and currently exist a Python toolkit which is able to implement feature selection methods along with subsequent classification results and metrics to evaluate the efficacy of these algorithms.

1.3   Contribution

We present a Python toolkit named *Chameleon*, which streamlines the process of feature selection for classification and evaluates the classification performances. Our toolkit implements six existing feature selection methods and an ensemble method of using features common to the methods evaluated. Six classifiers were also implemented, along with two classification performance metrics.

We evaluated the methods using four biological data sets and showed that although the classification performance varies according to data sets, feature selection and classification algorithms, the common features selected in general give better classification performance with less number of features.

2   Methods

2.1   Chameleon Toolkit Structure

Chameleon requires the installation of Python 3.6+ as well as Java. Currently, data sets to be used with the package must contain features and targets with either .mat or .arff formats. The chosen data set is first imported with the command **chameleon-data**. This command contains two sub-commands **format** and k-fold, which formats the data and partitions it into k folds to make k train and test data sets respectively. The value of k is set to 5, data normalised and the k-fold method is stratified by default, however these parameters may be adjusted by the user.

The full pipeline (Fig. 1) may then be configured using **chameleon-pipe**, whereby all feature selection methods and classifiers are added. Specific feature selection methods and classifiers may also be chosen and added individually. Finally, **chameleon-run** is used to run the configured pipeline. Feature selection methods are applied to the training data and the number of determined important features **n_features**, set to 50 by default, are used to build classification models. Subsequently, the test data is applied to the classification models in order

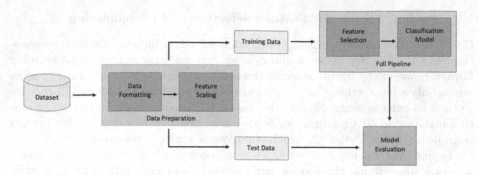

Fig. 1. General structure of Chameleon python toolkit.

to predict class labels, and classification evaluation metrics such as classification accuracy are able to be determined.

2.2 Feature Selection Methods

We included in *Chameleon* six feature selection algorithms and also implemented an ensemble method for feature selection:

Fisher Score. The aim of the Fisher Score (FS) feature selection algorithm is to select a subset of features such that the Euclidean distances between data points in differing classes are maximised and the distances between data points within classes are minimised [24]. The FS for each feature is calculated and the features are ranked according to their scores, with these highly ranked features being selected. Features are evaluated independent of each other, thus the FS method is less effective in handling feature redundancy as it does not evaluate combinations of features. Despite this, it has been shown to be effective in various applications, such as improving classification performance with medical data.

ReliefF. In the Relief feature selection algorithm, a feature statistic is calculated in order to help determine that feature's relevance to the classification problem [18]. The ReliefF algorithm improves the original Relief algorithm by iterating through a given number of training instances, and the feature statistics or scores are determined by feature value differences between nearest neighbour instances [19]. Thus, the Relief algorithm is able to determine relevant features in the context of others and take interactions between features into account.

Random Forest Feature Importance (RFFI). Random Forest (RF) classifiers train multiple decision trees on sub-samples, which are then aggregated for prediction. Such a classifier has shown to be effectively utilised as a method of feature selection [9]. When used in feature selection, Random Forests (RFs) were iteratively fitted to the data, with the least important features being discarded upon the creation of a new forest with each iteration. A RF feature importance algorithm is effective in obtaining small subsets of important features while maintaining high prediction accuracy.

SVM-Recursive Feature Elimination (SVM-RFE). Support Vector Machine (SVM) classifiers has been used with Recursive Feature Elimination (RFE) in order to select a small subset of features [16]. Here, a SVM classifier is used to train a model, where a decision boundary is determined in order to maximise the margin between classes, allowing for the weights of all the features to be determined accordingly. The algorithm then iteratively removes features with the smallest weights until a subset of relevant and non-redundant features remain, allowing for greater accuracy in the subsequent classification performance.

Iterative Mutual Information Selection. In the case of classification problems, Mutual Information (MI) refers to the amount that the knowledge provided by features that reduces uncertainty about the class. An approach to feature selection proposed by Battiti [3] considers the use of MI in order to evaluate features based on their relation to previously selected features and the class. The aim is to maximise the MI between the selected subset of features and target class, thus providing a subset which contains the most relevant and non-redundant information of the class. With a large set of features, such an algorithm can provide a significantly larger set of possible feature combinations to evaluate, increasing computational cost significantly. In iterative MI method, we select the subsequent feature based on the features already selected and the conditional MI score. A significance test is also run to ensure the new feature does add significantly to the conditional MI value, otherwise a feature with the highest MI value will be added.

Simple Mutual Information Score. In order to select a subset of optimal features using mutual information at a significantly lower computation cost than Battiti's algorithm, a simplified approach has been proposed in which information provided by features about the class are considered independent of each other [25]. The selected features are those with only the greatest amount mutual information with the class, taking into account features relevant to the classification problem but not allowing for the consideration of redundant features. As such, the results provided by the algorithm may be generally less accurate than alternative MI approaches, however such an approach benefits from significantly lowered computational expense.

Algorithm 1: Common Features Selection Method

---

    input : Set of feature ranking arrays S
                Size of feature subsets m
    output: Common features subset list f

1 $R \longleftarrow \emptyset$;
2 **for** $u \in S$ **do**
3     $\lfloor \;\; R \longleftarrow R \cup \{u[:m]\}$
4 $f = R_1 \cap \ldots \cap R_n$

2.3 Feature Selection with Common Features

It has been shown that ensemble methods can improve individual feature selection outcomes [4]. We therefore implemented in *Chameleon* an approach to select an optimal subset of features that are the common features among subsets selected by several high performing feature selection algorithms. The high performing feature selection algorithms are identified by users from classification results of individual feature selection methods. The proposed approach to selecting subsets of features common to various high-performing feature selection methods is described in Algorithm 1. Feature subsets of size m are determined from lists of features ranked by importance in descending order for each feature selection algorithm. The intersection between these subsets is found, resulting in a reduced subset of highly important features.

2.4 Classification Methods

Six common classification methods were implemented in *Chameleon*.

Gaussian Naive Bayes. The Gaussian Naïve Bayes (GNB) classifier is an algorithm which is based on Bayes' rule under the assumption of a Gaussian distribution in order to classify data [27]. Naïve Bayes (NB) algorithms operate under an assumption of conditional independence, which determines features to be independent of each other given the class variable.

K-Nearest Neighbours. K-nearest neighbours (KNN) classification allows for samples to be classified according to the class of their k-nearest neighbours [2]. The default values for *Scikit-Learn* Library's [22] KNN classifier were implemented, including a K value of 5 and Euclidean distance to measure the distance between points.

Logistic Regression. Logistic regression is a linear classifier that make use of the sigmoid function in order to classify data with binary target variables [10]. Logistic regression classifiers models the output probability that the input belongs to a certain class, transformed through the use of a logistic sigmoid function. The classifier allows for highly efficient training and high accuracy for linearly separable data sets. A significant limitation of the method is its assumption that the features and target variable share a linear relationship, as well as its tendency to over-fit when used with data sets containing a greater number of features than observations.

Neural Network (Multi-layer Perceptron). Multi-layer Perceptron (MLP) classifiers implement an underlying Neural Network (NN) in order to effectively classify data [23]. The parameters of the MLP classifier used include a limited-memory Broyden-Fletcher-Goldfarb-Shanno (lbfgs) solver to optimise the log-loss function, as well as hidden layer sizes of 100.

Random Forest. The RF algorithm builds a classifier using multiple decision tree classifiers [5]. The decision trees are trained using various random sub-samples and the results are subsequently aggregated for prediction of classes. The random forest algorithm uses bootstrap aggregation as well as random variable selection in order to build trees.

Support Vector Machine. SVM algorithms work to classify data by finding an optimal separating hyperplane between classes. A regularisation parameter C is determined, a lower value ensuring that the distance of groups from the margin is maximised, while a higher C value minimises the amount of misclassifications. The default values for *Scikit-Learn* Library's SVM classifier were used, including a C value of 1, and a Radial basis function (RBF) kernel in order to account for non-linearity that may be found in real-world data sets.

2.5   Performance Metrics

Two performance metrices were implemented in *Chameleon*: accuracy and Area Under the ROC Curve (AUC). Classification accuracy is the proportion of correct results to the total number of instances. AUC is a value between 0 and 1 that give indication on how effectively the classifier is able to separate the data into differing classes. Statistical significance levels for classification are also calculated through either a t-test for binary classification, or randomly permuting the test data labels for multi-class classification.

2.6   Data Sets

Four high-dimensional microarray gene expression data sets (Table 1) were used to evaluate the feature selection methods. The data sets represent both discrete and continuous features, as well as binary and multiple class targets.

Table 1. Details of chosen high-dimensional gene expression data sets

Data set	Instances	Features	Feature type	Classes	Reference
Colon cancer	62	2,000	Discrete	2	[1]
Leukemia cancer	72	7,129	Continuous	2	[14]
Glioma tumour	85	22,283	Continuous	2	[12]
Lung cancer	203	3,312	Continuous	5	[17]

2.7   Feature Selection Evaluation

Stratified 10-fold cross-validation is implemented for all data sets, partitioning the data into training and test sets, whilst preserving the proportions of class labels for each fold. Measures of classification accuracy for the six toolkit feature selection methods are computed, with models built using feature subsets with sizes ranging from 1 to 100 of the most important ranked features.

Classification accuracy of the common features selection method is determined by taking the common features apparent from feature subsets of multiple high-performing feature selection methods. The size of the feature subsets from which the common features are selected range from 10 to 100 included features, increasing with increments of 5.

We also evaluated the classification performance over the whole data set to provide a baseline for performance of selected features.

2.8   Software

This work uses the Python packages *Scikit-Feature* for four of the feature selection methods (RF, ReliefF, RFFI and SVM-RFE) [20] and *Scikit-Learn* for classifiers. The MI feature selection methods used the JIDT package [21] for MI calculations, and Python wrapper code is written to call methods in the package. The ensemble methods were developed and implemented by us in Python. The code used in this work is on GitHub[1].

3   Results and Discussion

3.1   Comparing Toolkit Feature Selection Methods

The feature selection algorithms included in *Chameleon* were applied to the four data sets in order to compute lists of feature importance. Classifications were then performed on the selected subsets and their performances compared.

Example results from two of the toolkit classifiers are shown for each data set, as all methods of classification yielded similar results. Classification accuracy results across all classifiers using the subset of first 100 selected features are

---

[1] https://github.com/diviyat/chameleon.

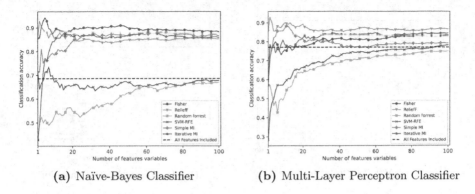

(a) Naïve-Bayes Classifier   (b) Multi-Layer Perceptron Classifier

Fig. 2. Colon cancer data set classification accuracy scores for six feature selection methods

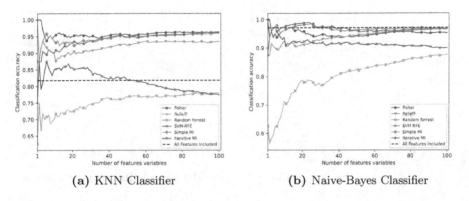

(a) KNN Classifier   (b) Naive-Bayes Classifier

Fig. 3. Leukemia cancer data set classification accuracy scores for six feature selection methods

shown in Tables 2, 3 and 4. For the colon cancer data set, we show results from NB and MLP classifiers (Fig. 2). For both classifiers, four of the feature selection methods (FS, ReliefF, and both MI methods) picked subsets that performed better than the baseline accuracy of using all features. The FS algorithm provides the most relevant features and the best accuracy for the NB classifier, while the ReliefF algorithm provides the highest performing feature subsets for the MLP classifier. The same features selection methods also provided high classification accuracies from KNN and NB for the Leukemia cancer data set (Fig. 3).

For the much higher dimensional Glioma brain tumour data set, three feature selection methods (FS and both MI methods) provided classification accuracy values that are better, or closely approach, the baseline accuracy (Fig. 4). A model fit to a SVM classifier with a subset of 40 features chosen using the iterative MI selection algorithm can be seen to provide better accuracy compared to a model built using all 22,283 features in the data set.

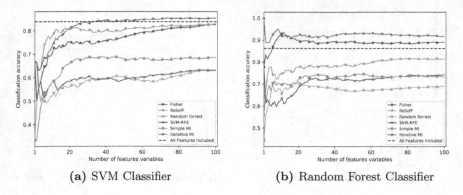

(a) SVM Classifier   **(b)** Random Forest Classifier

Fig. 4. Glioma brain tumour data set classification accuracy scores for six feature selection methods

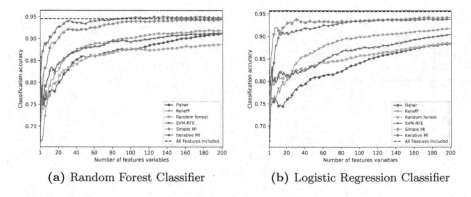

(a) Random Forest Classifier   **(b)** Logistic Regression Classifier

Fig. 5. Lung cancer data set classification accuracy scores for six feature selection methods

Finally, for the lung cancer data set, both MI methods can be seen to select highly relevant smaller feature subsets which result in accuracy values which out-perform or closely approach the baseline accuracy values for random forest and logistic regression classifiers (Fig. 5).

The results confirm that different feature selection methods do not perform uniformly. Certain feature selection algorithms included in the toolkit are able to better select a smaller subset of features that gives similar or improved classification results compared to classification accuracy computed using all features in a high-dimensional data set. Feature selection by MI methods can be seen be classifier agnostic as the feature subsets they selected result in generally high classification performance across the six different classifiers. Other feature selection algorithms provide the best classification results for certain classifiers, such as the ReliefF algorithm used to select a subset of features from the colon cancer data set for the MLP classifier (Fig. 2(b)).

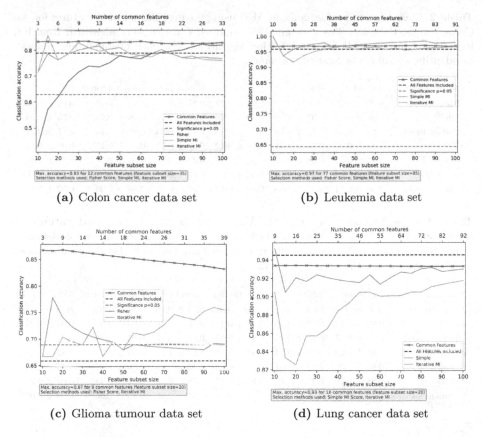

(a) Colon cancer data set

(b) Leukemia data set

(c) Glioma tumour data set

(d) Lung cancer data set

Fig. 6. Classification accuracy for common features selection method using a neural network (multi-layer perceptron) classifier.

3.2   Evaluating Common Features Selection Method

We show here the classification results from MLP using the common features for all data sets (Fig. 6). Features from SVM-RFE, RF or ReliefF were not included, as in general these features did not give good classification performance. The results show that while the size of the common feature set is much smaller than the respective feature sets they were chosen from, yet the classification performance is almost always better. Further, the common features achieve a higher classification performance at a much smaller set size than the features selected from the individual methods.

We compared the classification performance of all classifiers using the first 100 selected features from all feature selection methods and their common features subsets for colon cancer (Table 2), Glioma brain tumour (Table 3) and lung cancer (Table 4) data sets. 10-fold cross-validation evaluation was performed on the four high-dimensional data sets to gather the AUC values and the variance in the performance.

Table 2. Classification performance in AUC using the first 100 selected features for the Colon Cancer data set. Twenty-eight (28) common features were identified from FS, ReliefF and iterative MI feature selection algorithms. For each classification method, we identify, in italic, the best performance.

Feature selection	Naive-Bayes	KNN	Logistic reg	Neural net	Random forest	SVM
Fisher score	*0.93 ± 0.11*	*0.93 ± 0.10*	0.91 ± 0.11	0.86 ± 0.15	0.92 ± 0.10	0.92 ± 0.12
ReliefF	0.91 ± 0.11	0.92 ± 0.09	0.90 ± 0.09	0.89 ± 0.17	0.90 ± 0.13	0.95 ± 0.08
Random forest	0.69 ± 0.32	0.74 ± 0.19	0.77 ± 0.19	0.77 ± 0.24	0.82 ± 0.18	0.81 ± 0.21
SVM-RFE	0.74 ± 0.27	0.85 ± 0.13	0.90 ± 0.12	0.89 ± 0.12	0.90 ± 0.15	0.89 ± 0.17
Iterative MI	0.82 ± 0.21	0.88 ± 0.17	0.94 ± 0.17	*0.89 ± 0.11*	0.91 ± 0.11	0.90 ± 0.15
Simple MI	0.90 ± 0.10	0.91 ± 0.15	*0.94 ± 0.08*	0.88 ± 0.12	*0.93 ± 0.07*	*0.95 ± 0.11*
Common features	0.95 ± 0.11	0.92 ± 0.09	0.89 ± 0.16	0.86 ± 0.15	0.94 ± 0.08	0.92 ± 0.13

Table 3. Classification performance in AUC using the first 100 selected features for the Glioma Brain Tumour data set. Thirty-nine (39) common features were identified from FS, and iterative MI feature selection algorithms. For each classification method, we identify, in italic, the best performance.

Feature selection	Naive-Bayes	KNN	Logistic reg	Neural net	Random forest	SVM
Fisher score	*0.97 ± 0.04*	0.93 ± 0.10	*0.95 ± 0.08*	0.85 ± 0.15	*0.96 ± 0.06*	*0.97 ± 0.04*
ReliefF	0.89 ± 0.11	0.92 ± 0.10	0.91 ± 0.11	0.80 ± 0.19	0.96 ± 0.07	0.94 ± 0.05
Random forest	0.77 ± 0.15	0.79 ± 0.13	0.84 ± 0.13	0.81 ± 0.15	0.95 ± 0.08	0.77 ± 0.21
SVM-RFE	0.65 ± 0.22	0.76 ± 0.12	0.87 ± 0.10	0.67 ± 0.17	0.94 ± 0.08	0.89 ± 0.12
Iterative MI	0.94 ± 0.08	*0.95 ± 0.07*	0.94 ± 0.12	*0.85 ± 0.10*	0.96 ± 0.07	*0.97 ± 0.04*
Simple MI	0.81 ± 0.19	0.85 ± 0.17	0.80 ± 0.17	0.78 ± 0.18	0.86 ± 0.13	0.86 ± 0.07
Common features	0.96 ± 0.04	0.90 ± 0.16	0.86 ± 0.11	0.81 ± 0.13	0.96 ± 0.06	0.99 ± 0.02

For the colon cancer data set (Table 2), the feature selection algorithms determined to provide the best performing were FS, ReliefF and iterative MI algorithms. Using these methods to generate subsets including 100 of the most important features, 28 features were identified as common between them. It can be seen that the common features provide similar (within the margin of error) classification performance for all classifiers.

For the Glioma brain tumour data set (Table 3), the best performing feature selection algorithms were FS and iterative MI algorithms, which generated a subset with 39 features using the common features method. Classification results show the reduced sized common features subset has similar performance against for all classifiers except Logistic Regression.

Using both MI features selections, 88 common features were selected from 100 feature subsets for the lung cancer data set (Table 4). The classification performance using the common features were better than using the subsets from individual feature selection methods for all classifiers.

Table 4. Classification performance in AUC using the first 100 selected features for the Lung Cancer data set. Eighty-eight (88) common features were identified from iterative and simple MI feature selection algorithms. For each classification method, we identify, in italic, the best performance.

Feature selection	Naive-Bayes	KNN	Logistic reg	Neural net	Random forest	SVM
Fisher score	0.83 ± 0.11	0.94 ± 0.06	0.97 ± 0.02	0.95 ± 0.03	0.97 ± 0.03	0.96 ± 0.03
ReliefF	0.94 ± 0.04	0.97 ± 0.03	0.97 ± 0.02	0.95 ± 0.05	0.98 ± 0.02	*0.99 ± 0.01*
Random forest	0.95 ± 0.02	0.95 ± 0.04	0.96 ± 0.02	0.93 ± 0.04	0.97 ± 0.04	0.97 ± 0.02
SVM-RFE	0.90 ± 0.04	0.96 ± 0.02	0.96 ± 0.02	0.96 ± 0.04	0.97 ± 0.03	0.97 ± 0.02
Iterative MI	0.98 ± 0.02	*0.98 ± 0.02*	0.98 ± 0.02	*0.97 ± 0.02*	*0.99 ± 0.01*	*0.99 ± 0.01*
Simple MI	*0.99 ± 0.01*	0.97 ± 0.04	*0.98 ± 0.01*	0.96 ± 0.02	*0.99 ± 0.01*	*0.99 ± 0.01*
Common features	0.99 ± 0.01	0.98 ± 0.03	0.98 ± 0.02	0.97 ± 0.03	0.99 ± 0.01	0.99 ± 0.02

It can be seen that by utilising the best performing feature selection algorithms to select the features common between their determined feature subsets, improved or similar classification results are able to be achieved with smaller, and thus more computationally efficient, subsets of relevant and non-redundant features.

4   Conclusion

We presented *Chameleon*, a Python based toolkit for feature selection based on categorical variables and subsequent classification performance evaluation. *Chameleon* integrates various necessary steps including data pre-processing, k-fold cross validation, six feature selection algorithms, six different well-established classification methods, two classification performance metrics, as well as easy to understand visualisation of the results.

A rigorous analysis was conducted by means of combining the six classifiers with the six toolkit feature selection algorithms, as well as the introduced common features method, for binary and multi-class high-dimensional data sets. It was found that certain feature selection algorithms were able to select smaller subsets of features which yielded better classification performance compared to baseline accuracy computed with all features in the data sets. Using such high performing algorithms to select the common features between them resulted in improved or similar classification performance, compared to the individual methods of feature selection, using significantly smaller subsets of features.

It can be seen that the high-dimensional data sets tested result in generally high classification accuracy results. In future work, the toolkit's feature selection algorithms, as well as the proposed common features method, may be tested using more challenging data sets that are not so easily classified. Future development of *Chameleon* will include extending to more feature selection algorithms, adding other performance metrics, and extending the data formats permitted by the package beyond .mat and .arff formats.

Acknowledgement. Simon McManis was supported by 2019/2020 CSIRO Data61 Summer Scholarship for this project.

References

1. Alon, U., et al.: Broad patterns of gene expression revealed by clustering analysis of tumor and normal colon tissues probed by oligonucleotide arrays. Proc. Nat. Acad. Sci. **96**(12), 6745–6750 (1999)
2. Altman, N.S.: An introduction to kernel and nearest-neighbor nonparametric regression. Am. Stat. **46**(3), 175–185 (1992)
3. Battiti, R.: Using mutual information for selecting features in supervised neural net learning. IEEE Trans. Neural Netw. **5**(4), 537–550 (1994)
4. Bolón-Canedo, V., Alonso-Betanzos, A.: Ensembles for feature selection: a review and future trends. Inform. Fus. **52**, 1–12 (2019). https://doi.org/10.1016/j.inffus.2018.11.008
5. Breiman, L.: Random forests. Mach. Learn. **45**(1), 5–32 (2001)
6. Caruana, R., Freitag, D.: Greedy attribute selection. In: Machine Learning Proceedings 1994, pp. 28–36. Elsevier (1994)
7. Dash, M., Choi, K., Scheuermann, P., Liu, H.: Feature selection for clustering-a filter solution. In: Proceedings of the 2002 IEEE International Conference on Data Mining, pp. 115–122. IEEE (2002)
8. Dash, M., Liu, H.: Feature selection for classification. Intell. Data Anal. **1**(3), 131–156 (1997)
9. Díaz-Uriarte, R., De Andres, S.A.: Gene selection and classification of microarray data using random forest. BMC Bioinform. **7**(1), 3 (2006)
10. Dreiseitl, S., Ohno-Machado, L.: Logistic regression and artificial neural network classification models: a methodology review. J. Biomed. Inform. **35**(5–6), 352–359 (2002)
11. Dy, J.G., Brodley, C.E.: Feature selection for unsupervised learning. J. Mach. Learn. Res. **5**(Aug), 845–889 (2004)
12. Freije, W.A., et al.: Gene expression profiling of gliomas strongly predicts survival. Can. Res. **64**(18), 6503–6510 (2004)
13. Friedman, J., Hastie, T., Tibshirani, R.: The elements of statistical learning, vol. 1. Springer series in statistics New York (2001). https://doi.org/10.1007/978-0-387-84858-7
14. Golub, T.R., et al.: Molecular classification of cancer: class discovery and class prediction by gene expression monitoring. Science **286**(5439), 531–537 (1999)
15. Guyon, I., Elisseeff, A.: An introduction to variable and feature selection. J. Mach. Lear. Res. **3**(Mar), 1157–1182 (2003)
16. Guyon, I., Weston, J., Barnhill, S., Vapnik, V.: Gene selection for cancer classification using support vector machines. Mach. Learn. **46**(1–3), 389–422 (2002)
17. Hong, Z.Q., Yang, J.Y.: Optimal discriminant plane for a small number of samples and design method of classifier on the plane. Patt. Recogn. **24**(4), 317–324 (1991)
18. Kira, K., Rendell, L.A., et al.: The feature selection problem: traditional methods and a new algorithm. In: AAAI, vol. 2, pp. 129–134 (1992)
19. Kononenko, I., Šimec, E., Robnik-Šikonja, M.: Overcoming the myopia of inductive learning algorithms with RELIEFF. Appl. Intell. **7**(1), 39–55 (1997)
20. Li, J., et al.: Feature selection: a data perspective. ACM Comput. Surv. (CSUR) **50**(6), 94 (2018)

21. Lizier, J.T.: JIDT: an information-theoretic toolkit for studying the dynamics of complex systems. Front. Robot. AI **1**, 11 (2014). https://doi.org/10.3389/frobt. 2014.00011
22. Pedregosa, F., et al.: Scikit-learn: machine learning in Python. J. Mach. Learn. Res. **12**, 2825–2830 (2011)
23. Ruck, D.W., Rogers, S.K., Kabrisky, M.: Feature selection using a multilayer perceptron. J. Neural Netw. Comput. **2**(2), 40–48 (1990)
24. Sun, L., Zhang, X.Y., Qian, Y.H., Xu, J.C., Zhang, S.G., Tian, Y.: Joint neighborhood entropy-based gene selection method with fisher score for tumor classification. Appl. Intell. **49**(4), 1245–1259 (2019)
25. Wang, X.R., Lizier, J.T., Nowotny, T., Berna, A.Z., Prokopenko, M., Trowell, S.C.: Feature selection for chemical sensor arrays using mutual information. PLoS ONE **9**(3), e89840 (2014). https://doi.org/10.1371/journal.pone.0089840
26. Yu, L., Liu, H.: Efficient feature selection via analysis of relevance and redundancy. J. Mach. Learn. Res. **5**(Oct), 1205–1224 (2004)
27. Zhang, H.: The optimality of Naive Bayes, flairs conference (2004)

PostMatch: A Framework for Efficient Address Matching

Darren Yates[1]([✉]), Md Zahidul Islam[1], Yanchang Zhao[2], Richi Nayak[3],
Vladimir Estivill-Castro[4], and Salil Kanhere[5]

[1] School of Computing and Mathematics, Charles Sturt University,
Bathurst, Australia
{dyates,zislam}@csu.edu.au
[2] Data61, CSIRO, Canberra, ACT 2601, Australia
yanchang.zhao@csiro.au
[3] School of Electrical Engineering and Computer Science,
Queensland University of Technology, Brisbane, Australia
r.nayak@qut.edu.au
[4] Universitat Pompeu Fabra, Barcelona, Spain
vladimir.estivill@upf.edu
[5] University of New South Wales, Sydney, NSW 2052, Australia
salil.kanhere@unsw.edu.au

Abstract. Matching lists of addresses is an increasingly common task executed by business and governments alike. However, due to security issues, this task cannot always be performed using cloud computing. Moreover, addresses can arrive with spelling errors that can cause non-matches or 'false negatives' to occur. Our proposed framework, PostMatch, provides a locally-executed method for address-matching that combines the open-source 'Libpostal' address-parsing library with our 'postparse' post-processor code and machine-learning. PostMatch provides improved parsing accuracy compared with Libpostal alone, approaching 96.9%. The matching process features the Jaro-Winkler edit distance algorithm together with XGBoost machine-learning to achieve very high accuracy on public data. PostMatch is open-source (GPL3 licensed) and available as R script code on Github.

Keywords: Address matching · Data matching · XGBoost

1 Introduction

The need to match two lists of addresses can be a common occurrence in many organisations. For example, it could be to link or merge two customer databases from different retail store records or to identify fraud by matching addresses entered into application forms with those on a known 'black list'. However, address matching is often more complex due to the nature of the data itself. Address components, such as street name and town/suburb name, can be swapped or misaligned. This complexity can be the result of clients or customers

© Springer Nature Singapore Pte Ltd. 2021
Y. Xu et al. (Eds.): AusDM 2021, CCIS 1504, pp. 136–151, 2021.
https://doi.org/10.1007/978-981-16-8531-6_10

from different regional, cultural and/or language backgrounds entering address components differently into the same form. Added complexity can also arise from address components that have been misspelt or are incomplete. Moreover, the larger the address lists that are to be matched, the more system resources and search performance become a concern. Cloud computing services can be used to overcome many of these issues, however, there are many applications, such as law enforcement, where a locally-executed solution is required. This local-processing requirement could be due to the nature of the address lists or cyber security concerns with using online resources. As a result, matching the addresses from two large lists both accurately and efficiently can create significant challenges.

Other methods have been suggested previously to tackle some of these issues. For example, Christen and Belacic [6] developed an automated probabilistic solution employing Hidden Markov Models (HMMs) to guide the normalisation and parsing process. Address matching is a specific application of 'record linkage' theory pioneered by Fellegi and Sunter [8]. More recently, Koumarelas, Kroschk, Mosley and Naumann [11] investigated methods for combining address geocoding with a similarity measure to maximise address matching accuracy.

This paper introduces a machine-learning based address-matching framework called 'PostMatch'[1] to address these issues using locally-executed methods. The framework, shown in Fig. 1, consists of three key sections: parsing, normalisation and matching. The parsing process identifies and separates various address components, normalisation modifies those address components to fit in with agreed standard identifiers and features, and matching detects addresses common to both lists.

The research contributions of this paper include:

- identification of eight key address fields that summarise Australian addresses,
- a post-processing method called 'postparse' that improves the accuracy of the Libpostal open-source address parsing and normalisation library,
- the combination of machine-learning and string edit-distance measures to deliver high address-matching accuracy, and
- experimental evaluation of the proposed framework on public address data.

Although Australia is the example focus of this work, the methods used here could be applied to other national address formats following similar principles.

2   Related Work

Modern address matching can be traced back to work into 'record linkage' theory by Fellegi and Sunter [8], who applied a mathematical approach to the problem of matching two records using the records' components by comparing them as a vector of element pairs. The approach determined three possible matching labels. The first two are 'link' (indicating the two records match) and 'non-link' (do not match). Both of these were considered as definitive or 'positive' decisions. The

---

[1] https://github.com/darrenyatesau/postmatch.

third level, 'possible link', occurs when the match fails to meet the preconditions for either of the two positive outcomes. The aim is to increase the likelihood of a 'link' or 'no link' result, whilst minimising a 'possible link' outcome, which would require more expensive resources to link manually.

While record linkage has been well researched since then [9,10,14,16], Fellegi and Sunter identified initially that the accuracy of record matching would be directly affected by the completeness and accuracy of the initial records [8]. Within the specific area of address matching, these two issues noted by Fellegi and Sunter appear as missing or misspelt address fields. Thus, research has focused on achieving high matching accuracy in applications where address data is incomplete or contains errors that would ordinarily compromise that accuracy.

Christen and Belacic [6] tackled these issues of address cleaning and standardisation or 'normalisation' through hidden Markov Models (HMMs). They used this type of finite state machine in an attempt to identify various components from address strings. It also featured the Australian Geocoded-National Address File (G-NAF) database [6] as a verification source to determine that addresses being matched were genuine. More broadly, their approach features a four-step process involving 1) address cleaning or normalisation, 2) tagging of address component fields, such as street, town, state and so on, 3) segmenting the tag lists into appropriate output fields using the HMM and 4) verification against the G-NAF database.

A popular open-source software solution for address parsing and normalisation on a global scale is 'Libpostal' [1]. This library executes within Linux-based computer systems and supports application programming interfaces (APIs) for several programming languages, including Python and R. Libpostal authors claim it can support addresses from around the world. The Libpostal library offers two essential functions: 1) normalisation, to standardise an address using commonly-agreed and region-specific descriptors and features, and 2) parsing, to separate the address into individual components.

Even with address records parsed and normalised, the issues remain of how to accurately determine whether a pair of addresses are 'matched', 'not matched' or 'maybe matched'. In particular, there has been notable research into quantitative methods for determining the extent of matching between pairs of text-based records or 'strings', with numerous string-similarity measures developed, including Levenshtein [12] and Jaro-Winkler [15]. String-similarity measures quantify the steps or 'cost' required to transform one text string to another and are commonly used in record linkage applications.

By contrast, Arasu, Ganti and Kaushik developed a more generalised alternative [2]. Their work builds upon a concept of 'similarity join', whereby two databases are tested by each combination of record pairs against a similarity measure function, with those pairs that exceed a preset threshold being recorded. They acknowledged that despite the availability of numerous similarity or 'distance' functions, no one measure excels in every application. Their solution expanded the similarity join into a more extensive set of joins featuring multiple similarity functions to boost accuracy.

Practical applications of similarity measures have included matching health records. Bell and Sethi [3] developed a matching framework that also covers parsing and normalisation. A crucial part of the normalisation process was incorporating the ethnic origins of patients, to account for how patients from different backgrounds enter their name details. Their record-matching process features a composite matching formula that included string similarity and phonetic matching.

Cohen, Ravikumar and Fienburg [7] experimentally tested a number of common string distance measures available in an open-source Java programming language toolkit called 'SecondString'. One of these distance measures tested was the Levenshtein string distance function [12]. In broad terms, the function counts up the number of character transitions to move from one string to another. These transition options include insertion, deletion and substitution. Unlike the standard Hamming distance measure, the Levenshtein function handles strings of different length, an important factor for address matching.

We also draw attention to other experimental results in [7]. A test of 11 datasets comparing a series of distance measures revealed that, on average, the standard Levenshtein distance measure performed lower than the other measures on ten of those datasets. However, the one dataset it did excel on among these tests was a synthetic 'Census' dataset, consisting of names and addresses. Moreover, the results were further improved by applying the Winkler variation [15] to the Levenshtein measure. This variation aims to take into account common prefixes in strings by weighting them more favourably.

A more common non-scaled variation of the Levenshtein measure is the Damerau-Levenshtein (DL) function [13], which, in addition to insertion, deletion and substitution, includes the option of character transposition, or swapping of two adjacent characters, as part of the edit distance calculation. Each transformation carries an equal penalty or weight of one and the series of transformations that converts a string to the other string with the fewest steps becomes the edit distance for those two strings. However, work by Christen [5] on matching personal names showed that while the DL measure generally performed well in testing, alternatives, including the Jaro-Winkler (JW) algorithm [15], outperformed DL, confirming earlier results [7]. The JW measure has similarities with DL in that it accounts for insertion, deletion and transposition on characters. However, it differs by taking into account the number of characters in common between the two strings in addition to the number of transpositions relative to the length of the longer string. The tests [5] show that not only did the JW algorithm outperform DL in terms of matching accuracy in all tests, JW also processed the same number of tests at approximately twice to three times the speed of DL.

Thus, as noted, there may well be numerous distance measures, but it is apparent that no one measure is perfect for every application. However, while this might appear to strengthen the case for incorporating multiple edit measures, such as a set-similarity join, to cover all eventualities, the extra cost in time required to process these measures all but ensures there can be no gains without losses.

The research reported here shows that machine-learning techniques can be used to either enhance or replace the traditional rule-based solutions that are commonly applied to record linkage. One of the more recent machine-learning developments is XGBoost [4]. It is a gradient tree-boosting algorithm boasting high scalability, making it well suited to tasks such as address matching, where the number of addresses required to be checked may be considerable. It also offers high performance in single-machine use. To our knowledge, no research has yet been published into the efficacy of XGBoost in address-matching applications. Moreover, the combination of distance measures and XGBoost machine-learning forms one of the novel contributions of our PostMatch framework, which we will now cover in depth in the next section.

3   The PostMatch Framework

Our proposed PostMatch framework for efficient address matching is shown in Fig. 1. It consists of three key sections: 1) parsing, using the Libpostal open-source library, 2) normalisation, provided by our 'postparse' post-processing method, and 3) list-matching, involving set-similarity calculations based on edit distance and machine-learning algorithms.

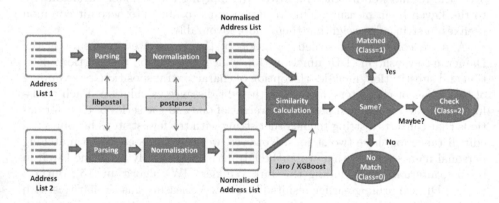

Fig. 1. The PostMatch address-matching framework

The framework begins with two separate address lists in the form of text files, where each address is a single string of text, one address per line. At first, the two lists are processed separately by parsing each address into component fields and then by normalising to reduce variations of values within fields. After that, address pairs from the two lists are compared using a combination of Jaro-Winkler edit-distance [15] and XGBoost machine-learning [4] algorithms. If an address pair is classed as 'matched' (class value of '1') or 'maybe-matched' (2), the class value and two addresses are combined to form a new record in the results matrix. After all address pair combinations have been tested, the results matrix can be displayed to the user.

3.1 'Site' and 'Locality' Fields in Addresses

With Australian postal addresses as a focus, our research has identified eight descriptor fields that can summarise all postal addresses within Australia's states and territories. These descriptors and their abbreviations are shown in Table 1. Moreover, these eight fields can be divided into two groups. The 'site' group, consisting of Fields 1 to 4, indicate an individual site, such as a building or subsection of a building (as in a level or unit/apartment, or even a post office box). The 'locality' group, featuring the four remaining fields, covers progressively larger groups of sites, from streets to towns and regions. A fundamental differentiator between the two groups is that while the 'site' group fields are, in some respects, optional (and generally mutually exclusive, in the case of PBox and HseNo fields in Australian addresses), the final three fields of the 'locality' group - Town, State and PCode - are essential to ensure accurate postal location identification. This can be seen in the three fictitious but standards-compliant Australian addresses examples shown in Fig. 2.

Table 1. The eight descriptor fields for identifying all Australian postal addresses, including abbreviations, content type and assigned group in our research.

Field no	Address descriptor	Abbrev'n	Content type	Group
1	Post Office (PO) Box	PBox	Alpha-numeric	Site
2	Apartment/Unit number	Unit	Alpha-numeric	Site
3	Building Level number	Level	Alpha-numeric	Site
4	House/Site number	HseNo	Alpha-numeric	Site
5	Street name	Street	Alpha-numeric	Locality
6	Suburb/Town name	Town	Alpha-numeric	Locality
7	State name	State	Alpha-numeric	Locality
8	Postal code	PCode	Numeric (4-digit)	Locality

3.2 Address Parsing

Addresses can be described as region-specific location descriptors that follow local conventions or customs. Thus, it is not only important to identify the key address components, but also to identify where those components occur within an address. The general order for Australian addresses is in increasing order of scale, that is, building or 'site', street, town, region/state. The 'postal code' field is usually associated with the 'town/suburb' field, though in Australia, these two

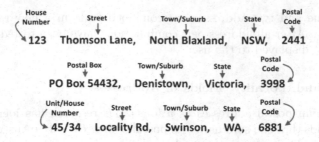

Fig. 2. An example of three (fictitious) compliant Australian-standard addresses. Each line in black is an address and the text in blue shows various fields in addresses.eps

do not always form exclusively distinct pairs and it is not uncommon for smaller localities to share a common postal code. By contrast, some towns may also have more than one post code. Examples of valid (albeit fictitious) addresses and their fields are shown in Fig. 2. The task of separating these components into identifiable fields is referred to as 'parsing'. The complexity of this task arises from the need to be able to recognise the various fields simply from the address format itself. The role of parsing within the PostMatch framework is to separate the address components appropriately into the eight component address fields (see Table 1), using Libpostal [1], an open-source address normalisation and parsing library.

3.3   Normalisation

Once the address fields have been identified and separated, each field has to be normalised for common extensions and abbreviations to improve matching accuracy. The normalisation process firstly reduces redundant information, such as 'PO Box 123' to simply the PO Box number itself. This removes the potential for post office prefixes to causes non-matches for matched fields. However, in Australia, normalisation should also handle a common forward-slash '/' abbreviation typically used to separate apartment/unit from building site descriptors. For example, '12/34 qwerty st' is a common short-form representing 'Unit 12, site or building number 34, Qwerty Street'. Another example is the 'street' field, where the word 'street' itself is often abbreviated to 'st', 'avenue' to 'ave' or 'av', and 'road' to 'rd'. Moreover, Australia's eight states also have common abbreviations, such as 'NSW' for New South Wales, 'QLD' for Queensland and so on. A common approach is to turn the name-based fields into lower-case and expand all abbreviations to their root form. This normalisation can be achieved through look-up tables or simple rule-based substitutions.

Thus, the task of normalisation is to process abbreviations and other common short-forms back into a set of consistent alphanumeric site descriptors to simplify the matching process and improve matching accuracy. This is the task of our 'postparse' post-processing method (see Sect. 3.4) that is applied to the address records after parsing is complete.

3.4   The 'Postparse' Post-Processing Method

While Libpostal is capable of handling a wide array of address formats from countries around the world, a number of issues were observed during testing with regards to incorrect normalisation of certain Australian-specific address components. For example, the Australian state abbreviation 'WA', commonly used for 'Western Australia', would expand to 'Washington', one of the United States. It also incorrectly handled addresses for Northern Territory, the Libpostal normalisation expanding the short-form 'NT' to 'Intermediary', as well as the 'street' reduction, 'st', to 'saint'. Further, it did not recognise the forward-slash '/' short-form for unit or apartment, instead replacing it with a space character. Similar issues with Libpostal have also been identified by Koumarelas et al. [11].

Our method for overcoming parsing and normalisation issues in Libpostal is to identify patterns of inconsistency and to correct them through a separate process after the event, commonly referred to as 'post-processing'. The PostMatch framework achieves this through an R script called 'postparse' that slots into the PostMatch framework following completion of address parsing by Libpostal.

In general, the traditional workflow process would be to use Libpostal to first normalise the address to ensure consistent field values, then parse the address into its component fields. However, due to the normalisation inconsistencies with Libpostal, experiments in Sect. 4 will show that it is possible to achieve greater levels of matching accuracy by using Libpostal as a parsing agent only and applying a targeted normalisation technique afterwards. As a result of those experiments, the 'postparse' post-processing method applies a number of corrective procedures to a parsed address, as follows.

Searching Libpostal 'suburb' and 'city' Output Fields. During our experiments on the parsing process, it was noted that Libpostal outputs the parsed Australian town/suburb value in either the 'suburb' or 'city' fields, but not in both. The results of experiments in Sect. 4 show that overall parsing accuracy is improved by searching both of these output fields for the PostMatch 'Town' field value, regardless of whether the value is misspelt or not.

Unit/Building Number Deciphering. On testing with addresses featuring the forward-slash unit/building number abbreviation, it was also noted that Libpostal's normalisation process would simply replace the forward-slash with a space, for example, '12/34' (i.e., unit 12 of building/house no. 34) would become '12 34' and this value would be incorrectly assigned to the Libpostal 'house_number' output field. However, in contrast, bypassing Libpostal normalisation for parsing only resulted in the original forward-slash notation value being assigned to the 'house_number' output field. As a result of this behaviour, 'postparse' checks the house-number field for a forward-slash character, splitting the values either side and assigning them to the appropriate fields. In this way, both unit number and house/building number can be identified correctly.

Misspelt Town Field-Parsing Correction. As will also be noted in Sect. 4, Libpostal loses accuracy when the 'town' name in an address is misspelt. Further, if the 'town' name is misspelt and contains two words, such as 'North Strathfield', Libpostal can mis-parse both the town and street fields, adding all but the last word of the 'town' name to the end of the 'street' field value. For example, '123 railway parade clareview heghts' would be parsed with {street = 'railway parade clareview'} and {suburb = 'heghts'}. The postparse post-processor handles this by determining if a known street suffix (e.g. street, parade, drive etc.) is the last word in the 'street' field value. If not, the 'street' field is searched for a known suffix, with all words following the suffix removed and added to the start of the 'suburb' field.

Normalising State Names. It has been observed that Libpostal can randomly alternate between the full Australian state names and their common abbreviations depending on the condition of the address, for example, 'victoria' becomes 'vic' or vice versa. However, in the many examples seen, the state is always correct, even if the value is not consistent between full and short names. Nevertheless, the difference between full and short names technically constitutes different strings and would be seen as 'not matched' by strict edit distance measures. To cater for this behaviour, PostMatch checks for inconsistencies in parsing and reverts any full state names back to their three-letter standardised abbreviations for Australian states and territories.

State Field Value Spelling Correction. Given that Australia only has eight states and territories, PostMatch provides spelling correction using a combination of n-grams and the Damerau-Levenshtein (DL) edit distance [13]. If Libpostal fails to provide a valid 'state' output value, the entire address string is searched as a series of n-grams (n = 1,2,3). The state name that has the lowest edit distance to one of the address n-grams is then chosen. The need to allow for values up to n = 3 in n-grams is to allow for the state names 'australian capital territory' and 'new south wales' as possible address 'state' field values. An example of this is shown in Fig. 3.

3.5   Address Pair Matching

Once the addresses have been parsed and normalised, they are ready to be match-checked. Taking a leaf from Fellegi and Sunter [8], PostMatch employs a combination of Jaro-Winkler edit distance [15] and XGBoost machine-learning [4] to compare addresses in two lists, identifying addresses that appear in both lists as 'matched' or 'maybe matched'. The process involves first taking an address pair to be checked and converting it into an edit difference record. This is done by taking a corresponding field from each address and calculating the edit distance or cost for transitioning from one field value to the other. The edit distance becomes the attribute value for that field pair in the new record. Thus, the eight fields of each address are combined to form eight edit-distance attributes

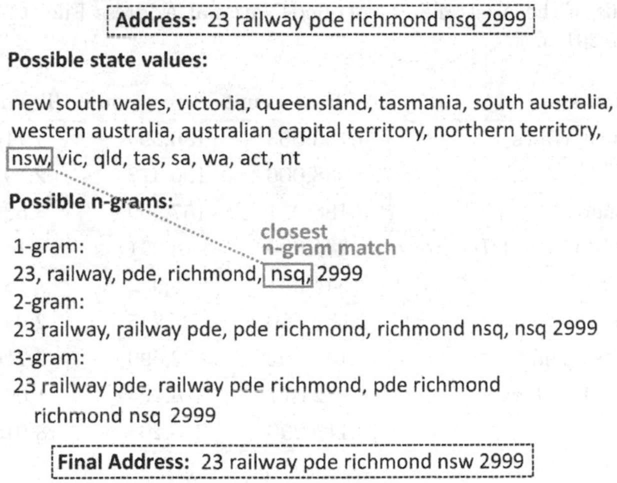

```
┌─────────────────────────────────────────────────┐
│ Address:  23 railway pde richmond nsq 2999        │
└─────────────────────────────────────────────────┘
```

Possible state values:

new south wales, victoria, queensland, tasmania, south australia,
western australia, australian capital territory, northern territory,
⟦nsw,⟧ vic, qld, tas, sa, wa, act, nt

Possible n-grams:

        closest
1-gram: n-gram match
23, railway, pde, richmond, ⟦nsq,⟧ 2999
2-gram:
23 railway, railway pde, pde richmond, richmond nsq, nsq 2999
3-gram:
23 railway pde, railway pde richmond, pde richmond
  richmond nsq 2999

```
┌─────────────────────────────────────────────────┐
│ Final Address:  23 railway pde richmond nsw 2999  │
└─────────────────────────────────────────────────┘
```

Fig. 3. Using n-grams and DL edit distance to correct state spelling errors.

in the new record. Once all eight address field pairs have been processed and the new record completed, it is tested against a machine-learning model previously learned from a training dataset of known address pair records. The predicted result of the model determines whether the current address pair is matched, not-matched or maybe-matched.

A training dataset containing class-labelled records is required to train the machine-learning model. In practice, this would be obtained through expert domain knowledge or from an existing reference dataset, depending on the application. However, as an existing labelled training dataset was not available during research, Sect. 4 will now describe the experimental phase, including the development of a synthetic training dataset from a public address data source.

4   Experimental Evaluation

4.1   Data and Experimental Environment

A public dataset, the Australian Geocoded-National Address File (G-NAF) [6], is used to evaluate the performance of our proposed PostMatch framework. G-NAF is an Australian government initiative to provide an open database of all physical addresses in all eight Australian states and territories[2]. It contains over 15 million addresses and Table 2 shows its statistics.

All experiments reported in this work were conducted on a desktop PC with a 3GHz Intel quad-core Core i5 CPU and 16GB of RAM, running Xubuntu 16.04.4 operating system and RStudio development environment. The Libpostal library was locally compiled from source code downloaded in March 2019.

---

[2] G-NAF: https://data.gov.au/data/dataset/19432f89-dc3a-4ef3-b943-5326ef1dbecc.

Table 2. Details of the Australian Geocoded-National Address File (G-NAF) version 201908 (August 2019).

Australian state	Address count	Street count	Town count
New South Wales	4,703,553	180,594	4,716
Victoria	3,838,006	190,472	2,997
Queensland	3,188,961	157,900	3,529
Australian Capital Territory	241,000	8,251	142
Tasmania	344,231	18,952	781
Western Australia	1,532,860	82,245	2,041
South Australia	1,153,802	72,590	2,716
Northern Territory	112,877	8,204	1,096
TOTAL	15,115,290	719,208	18,018

4.2   Experiments on Address Parsing

To begin, a set of complete addresses were derived from G-NAF, totalling a maximum of 200,000 for each state, for a total in excess of 1.5 million. From this, a test dataset of 100,000 addresses were randomly selected. Excess information was removed and each address was formed into a record of eight fields noted in Table 1. After that, eight fields of each address were concatenated into a single string of text. To test PostMatch's parsing ability in the face of addresses with spelling errors, we created a second dataset by injecting spelling errors, e.g., by swapping a vowel in town name.

We conducted a series of seven tests to compare PostMatch with various parsing and normalisation combinations of Libpostal. The tests are:

- Test 1: libpostal parsing. This involved using only Libpostal's parsing routine, with all suburb/town values being selected from Libpostal's 'Suburb' output field only.
- Test 2: libpostal parsing with 'town' search (libpostal-STC). During testing, Libpostal was identified to split parsing of suburbs and towns between its 'Suburb' and 'City' fields. This test is the same as Test 1, except that the final PostMatch 'town' value is selected from either Libpostal's 'Suburb' or 'City' output fields.
- Test 3: libpostal normalisation and parsing. This test first normalises each address using Libpostal's normalisation routine, then parses it with Libpostal. This test most mirrors standard Libpostal usage.
- Test 4: libpostal-State normalisation and parsing (libpostal-STATE). This test is the same as Test 3, but allows state values to be either in the full or short forms. This is to address the issue that, when normalising, Libpostal inconsistently converts some state abbreviations to their full names.
- Test 5: combines Tests 2 and 3
- Test 6: combines Tests 2 and 4

– Test 7: PostMatch (our method). This test runs our PostMatch framework, combining Libpostal parsing with 'postparse' post-processing.

Results on Addresses Without Spelling Errors. The results of the seven tests on the address dataset without spelling errors are shown in Table 3, where our proposed method is of the highest accuracy (highlighted in bold). As previously noted, common usage of Libpostal is to enact its normalisation process on an address first, then apply the results to its parsing engine. Comparing the results of Test 1 and Test 3, it appears that for Australian addresses at least, Libpostal works best when its normalisation process is not used and the addresses are parsed directly. Moreover, the 97.9% accuracy result of Test 2 for Libpostal-STC (searching 'suburb' and 'city' fields for the suburb value) shows that Libpostal is capable of high-accuracy parsing. However, applying Libpostal parsing followed by our 'postparse' post-processing improves the parsing accuracy further to 99.4%, albeit at the cost of 1:20 (i.e., 1 min 20 secs) of extra processing time (3:05 vs 1:45).

Table 3. Experimental results on address data with no spelling errors. The best method is highlighted in bold.

Test no	Parsing	Normalisation	Records matched (out of 100,000)	Percent matched (%)	Process time (mins:secs)
1	Libpostal	–	61,687	61.6	1:42
2	Libpostal-STC	–	97,957	97.9	1:45
3	Libpostal	Libpostal	31,966	31.9	2:08
4	Libpostal-STATE	Libpostal-STATE	47,076	47.0	N/A
5	Libpostal-STC	Libpostal-STC	47,810	47.8	2:09
6	Libpostal-STC-STATE	Libpostal-STC-STATE	72,958	72.9	N/A
7	**Libpostal**	**postparse (our method)**	**99,418**	**99.4**	**3:05**

Results on Addresses with Spelling Errors. The results of the seven tests on address data with spelling errors are shown in Table 4. These accuracy levels are well down compared with those shown in Table 3. Moreover, introducing the Libpostal normalisation before parsing mis-spelt addresses reduces parsing accuracy further still, with all results well below those of the correctly-spelt addresses tested in Table 3 and at best, reaching only 56.1%. The most successful technique tested is PostMatch, our combination of Libpostal parsing and 'postparse' post-processing, where parsing accuracy was largely maintained at just under 97% (Table 4, Test 7), a fall of less than 3% compared with the correctly-spelt result of Table 3. This is in comparison to the 'Libpostal-STC' results, where the fall

Table 4. Experimental results on address data with spelling errors. The best method is highlighted in bold.

Test no	Parsing	Normalisation	Records matched (out of 100,000)	Percent matched (%)	Process time (mins:secs)
1	Libpostal	–	13632	13.6	1:50
2	Libpostal-STC	–	74477	74.4	1:52
3	Libpostal	Libpostal	11388	11.3	2:19
4	Libpostal-STATE	Libpostal-STATE	12487	12.4	N/A
5	Libpostal-STC	Libpostal-STC	38303	38.3	2:19
6	Libpostal-STC-STATE	Libpostal-STC-STATE	56126	56.1	N/A
7	**Libpostal**	**postparse (our method)**	**96920**	**96.9**	**3:30**

Table 5. Experimental results on address data with field swaps. The best method is highlighted in bold.

Test no	Parsing	Normalisation	Records matched (out of 100,000)	Percent matched (%)	Process time (mins:secs)
1	Libpostal-STC	–	59030	59.0	1:49
2	Libpostal-STC	Libpostal-STC	26793	26.7	2:16
3	Libpostal-STC-STATE	Libpostal-STC-STATE	40896	40.8	N/A
4	**Libpostal**	**postparse (our method)**	**59833**	**59.8**	**6:02**

in accuracy from correctly-spelt (Table 3, 97.9%) to misspelt (Table 4, 74.4%) is in excess of 23%. Thus, the small fall of PostMatch from 99.4% to 96.9% shows the efficacy of this approach with addresses containing spelling errors.

Results on Addresses with Field Swaps. To test the robustness of different methods, we randomly changed the order of some fields within addresses and conducted another experiment. An example of field order change is to push the 'Town' field to the end of the string and move the 'State' and 'PCode' fields closer to the front of the string. The results are shown in Table 5, which suggests that in its current form, Libpostal (together with our postparse post-processing) is only able to correctly parse six out of ten Australian addresses (59.8%) at best. Moreover, PostMatch (Test 4) shows only a minor improvement in parsing accuracy, whilst processing time has increased significantly (6:02 vs 3:30 in Test 7, Table 4) as a result of the change in field order.

4.3   Experiments on Address Matching

Following the same approach used in Sect. 4.2, we created a dataset containing 200,000 address pairs for address matching evaluation. The address pairs are of three classes:

- Class '0': unmatched address pairs,
- Class '1': matched address pairs, and
- Class '2': matched address pairs with some fields changed using the same perturbation method used in Sect. 4.2, to test the model's ability of dealing with misspellings and field misalignment.

Table 6. Experimental results on address matching.

Predicted	Actual		
	Class = 0	Class = 1	Class = 2
Class = 0	33361	0	6
Class = 1	0	33300	0
Class = 2	0	55	33278
Sensitivity	1.0	0.9984	0.9998
Specifivity	0.9999	1.0	0.9992
Balanced accuracy	1.0	0.9992	0.9995

Each class covers $1/3$ of the data. We split the above data into training and test datasets in a 50:50 ratio.

With our PostMatch framework (see Fig. 1), each record of above data was first parsed with Libpostal and processed with our postparse processing method. After that, the Jaro-Winkler (JW) algorithm [15] was used to calculate the edit distance between the corresponding fields of each address pair. Finally, the edit distances of the eight address fields, together with the class label, were fed into a XGBoost [4] model for prediction.

Results of Address Matching. The results of address matching are shown in Table 6. The speed of the XGBoost algorithm allowed the 100,000 record-pairs to be processed in 12.91 ms. This would enable one million record-pairs to be processed every 130 ms, or 0.13 μs per record-pair. However, this is one million address comparisons required to compare two normalised addresses containing only 1,000 addresses each. Thus, since the computational load increases by $O(n^2)$, where n is the number of addresses per list, this task requires reasonable levels of computing power to execute locally. Still, the XGBoost matching component of this framework would be able to compare two normalised address lists, each containing 100,000 records, in less than 22 min on our test quad-core computer system.

5   Conclusions

This paper introduced the PostMatch framework for efficient address matching using a combination of the Libpostal open-source address parsing software and Jaro-Winkler/XGBoost algorithms. We have identified the issues of Libpostal and introduced a 'postparse' post-processing routine to improve it. Experimental results show that the combination of Libpostal and 'postparse' routines can achieve parsing accuracy of 99.4% with correctly-spelt addresses and even 96.9% with addresses featuring misspelt components. Our PostMatch framework built with a Jaro-Winkler similarity measure and an XGBoost model achieved near-100% matching accuracy. Nevertheless, we must also outline the areas for improvement, including greater understanding of Libpostal's ability to parse addresses with varying field order, as well as addresses with missing values.

The PostMatch framework is generic, although the proposed method was evaluated with Australia address data. We plan to confirm that our generic method can be adapted and applied to addresses from other counties or for other languages in future work.

References

1. Libpostal: international street address NLP, March 2019. https://github.com/openvenues/libpostal
2. Arasu, A., Ganti, V., Kaushik, R.: Efficient exact set-similarity joins. In: Proceedings of the 32nd International Conference on Very Large Data Bases, pp. 918–929. VLDB Endowment (2006)
3. Bell, G.B., Sethi, A.: Matching records in a national medical patient index. Association for Computing Machinery. Commun. ACM **44**(9), 83 (2001)
4. Chen, T., Guestrin, C.: XGBoost: a scalable tree boosting system. In: Proceedings of the 22nd ACM SIGKDD International Conference on Knowledge Discovery and Data Mining, pp. 785–794. ACM (2016)
5. Christen, P.: A comparison of personal name matching: techniques and practical issues. In: Sixth IEEE International Conference on Data Mining-Workshops (ICDMW 2006), pp. 290–294. IEEE (2006)
6. Christen, P., Belacic, D.: Automated probabilistic address standardisation and verification. In: Australasian Data Mining Conference (2005)
7. Cohen, W., Ravikumar, P., Fienberg, S.: A comparison of string metrics for matching names and records. In: KDD Workshop on Data Cleaning and Object Consolidation, vol. 3, pp. 73–78 (2003)
8. Fellegi, I.P., Sunter, A.B.: A theory for record linkage. J. Am. Stat. Assoc. **64**(328), 1183–1210 (1969)
9. Hand, D., Christen, P.: A note on using the f-measure for evaluating record linkage algorithms. Stat. Comput. **28**(3), 539–547 (2018)
10. Koudas, N., Sarawagi, S., Srivastava, D.: Record linkage: similarity measures and algorithms. In: Proceedings of the 2006 ACM SIGMOD International Conference on Management of Data, pp. 802–803. ACM (2006)
11. Koumarelas, I., Kroschk, A., Mosley, C., Naumann, F.: Experience: enhancing address matching with geocoding and similarity measure selection. J. Data Inf. Qual. (JDIQ) **10**(2), 8 (2018)

12. Levenshtein, V.I.: Binary codes capable of correcting deletions, insertions, and reversals. In: Soviet Physics Doklady, vol. 10, pp. 707–710 (1966)
13. Van der Loo, M.P.: The stringdist package for approximate string matching. R J. **6**(1), 111–122 (2014)
14. Vatsalan, D., Sehili, Z., Christen, P., Rahm, E.: Privacy-preserving record linkage for big data: current approaches and research challenges. In: Zomaya, A.Y., Sakr, S. (eds.) Handbook of Big Data Technologies, pp. 851–895. Springer, Cham (2017). https://doi.org/10.1007/978-3-319-49340-4_25
15. Winkler, W.E.: String comparator metrics and enhanced decision rules in the Fellegi-Sunter model of record linkage (1990)
16. Winkler, W.E.: Matching and record linkage. Bus. Survey Meth. **1**, 355–384 (1995)

Detection of Classical Cipher Types with Feature-Learning Approaches

Ernst Leierzopf[1]([⊠]), Vasily Mikhalev[2], Nils Kopal[2], Bernhard Esslinger[2], Harald Lampesberger[1], and Eckehard Hermann[1]

[1] University of Applied Sciences Upper Austria, Hagenberg, Austria
`ernst.leierzopf@ins.jku.at,`
`{Harald.Lampesberger,Eckehard.Hermann}@fh-hagenberg.at`
[2] University of Siegen, Siegen, Germany
`{Vasily.Mikhalev,Nils.Kopal,Bernhard.Esslinger}@uni-siegen.de`

Abstract. To break a ciphertext, as a first step, it is essential to identify the cipher used to produce the ciphertext. Cryptanalysis has acquired deep knowledge on cryptographic weaknesses of classical ciphers, and modern ciphers have been designed to circumvent these weaknesses. The American Cryptogram Association (ACA) standardized so-called classical ciphers, which had historical relevance up to World War II. Identifying these cipher types using machine learning has shown promising results, but the state of the art relies on engineered features based on cryptanalysis. To overcome this dependency on domain knowledge, we explore in this paper the applicability of the two feature-learning algorithms long short-term memory (LSTM) and Transformer, for 55 classical cipher types from ACA. To lower the necessary data and the training time, various transfer-learning scenarios are investigated. Over a dataset of 10 million ciphertexts with a text length of 100 characters, Transformer correctly identified 72.33% of the ciphers, which is a slightly worse result than the best feature-engineering approach. Furthermore, with an ensemble model of feature-engineering and feature-learning neural network types, 82.78% accuracy over the same dataset has been achieved, which is the best known result for this significant problem in the field of cryptanalysis.

Keywords: Cipher-type detection · Classical ciphers · Neural networks · Transfer learning · Ensemble learning

1 Introduction

In machine learning, classification is a problem of assigning data objects to the predefined categories by recognizing category-specific regularities in the data. There are various approaches of how neural networks can be applied to solve this problem. One way is to first manually preprocess the raw data and to calculate features which are then used by neural network for classification. In this paper, the term feature engineering (FE) is used when referring to this approach.

© Springer Nature Singapore Pte Ltd. 2021
Y. Xu et al. (Eds.): AusDM 2021, CCIS 1504, pp. 152–164, 2021.
https://doi.org/10.1007/978-981-16-8531-6_11

It can be very efficient in many cases as it allows to use the domain knowledge when constructing the features. Domain knowledge in the field of cipher type detection is based on statistical measures. For example it is known that transposition ciphers have similar frequency distributions as the English language, but with different letters. Substitution ciphers are also distinguishable, because of the different algorithms used for encryption. However, the feature-engineering approach struggles from two main drawbacks: i) It requires a long and costly process of defining and testing different features. ii) It only works for problems where domain knowledge is available.

Another approach is to use neural networks which have a built-in mechanism for detecting regularities in data. This approach is often referred to as feature learning (FL). Examples of feature-learning architectures are recurrent neural networks (RNN) like long short-term memory (LSTM) [6], and Transformer-based neural networks [19].

This paper compares FE and FL approaches applied to the problem of classification of (a large synthetic set of) classical ciphers, which were historically relevant up to World War II. Current research focuses on the automated digitization, analysis, and decryption of encrypted historical documents [15]. Ciphers (or encryption algorithms) are a method to protect information (plaintext) by converting it to ciphertext. Ideally, only someone who knows the secret (key) is able to decipher a ciphertext. Therefore, strong encryption algorithms are used to achieve a high level of security and information privacy. However, if a cipher has weaknesses, a cryptanalyst (or attacker) could break it and recover the original information from the ciphertext without knowledge of the key. If the attacker has access only to the ciphertext (ciphertext-only attack), the first step is to determine which cipher was used to produce it. Then further techniques are applied for recovering the information. For this reason, cipher-type classification is a crucial part of cryptanalysis. Note, that modern ciphers are designed in such a way that it is extremely difficult to distinguish their ciphertexts from randomly generated data [7]. Therefore, classification of modern ciphers is usually impossible with the FE approach. On the other hand, a FL approach can potentially detect regularities in ciphertexts which are not detected by common methods yet. As classical ciphers usually have many weaknesses, FE approaches can be very effective for their classification as we demonstrated in [13].

This work investigates how well FL approaches perform when applied to this cipher-type-detection problem. For this research, a large amount of ciphertexts was generated using 56 classes – 55 different classical ciphers standardized by the American Cryptogram Association (ACA)[2] and one plaintext class. Without any feature pre-computation, these ciphertexts were provided as raw data to the LSTM and Transformer neural networks to detect the cipher type used to produce them. Also ensemble models with FE-only approaches, FL-only approaches and combined FE and FL approaches, were evaluated.

We also evaluate whether transfer learning is applicable to this kind of classification problem. Our results show: If a feed-forward neural network (FFNN) is trained to classify a certain number of ciphers and if at a later step previously not included ciphers are provided to it, then the learning process becomes much faster compared to training the complete model again. In other words, transfer learning can be successfully applied to this problem and catastrophic forgetting does not happen.

To summarize, our key contributions are:

- We demonstrate the feasibility of the FL approaches to the problem of classical cipher-type detection.
- By using the combination of various techniques in an ensemble model, we achieve the accuracy of 82.78% which is the best state-of-the art-result for this problem.
- We show that transfer learning is applicable here, which means that the knowledge gained through our experiments can also be helpful for the different related problems.

2   Related Work

We described in [13] the state-of-the-art feature map for all ACA ciphers and the according hyperparameters for the FE approaches FFNN, random forests (RF) and naïve Bayes networks (NBN). As a result it is shown that RF needs the smallest amount of training data in order to achieve acceptable results (about 1.5% data of the FFNN). Different optimizers and activation functions were tested with the baseline FFNN. The best and most comparable results were achieved with the Adam optimizer and the ReLU activation function. Out of the total 27 different features, which were tested one by one, 20 were used for the final feature map [13].

All relevant related work in the field of classical cipher-type detection is listed in Table 1. All results in Table 1 are based on FE to classify cipher types. They are, however, not directly comparable, because every listed work uses their own set of cipher types, features and datasets. For example, in [1] a very high accuracy of 99.60% is shown. However, this result is achieved by using ciphertexts which are 1 million characters long. Such a length is unrealistic for any practical scenario of classical ciphers usage. In other works [9,10,18], the classification is done among a very small number of cipher types – from 3 to 6, as compared to 56 types considered in this research. The most notable related works are from Nuhn and Knight [17] and from the authors [13] who solve the classification problem for the comparable number of ciphers – 50 and 56 respectively. Nuhn and Knight achieved an accuracy of 58.50% for 50 ACA ciphers using a neural network with a linear classifier and a Stochastic Gradient Descendent (SGD) optimizer with default parameters. To accomplish this, they used 58 features and 1 million ciphertexts with random lengths in 20 epochs for training [17]. Besides neural networks, other technologies like Support Vector Machines (SVM), Hidden Markov Models (HMM), Decision Trees (DT) and multi-layer classifiers were used in the listed work. Multi-class classification in SVMs can only be achieved by using many one-vs-one binary classifiers. This is the reason for the long training times of the SVM and the reason why it is not implemented in our current solution.

Results from the work of Sivagurunathan et al. [18], where the three classical ciphers Playfair, Hill and Vigenère were analyzed with a simple neural network, coincide with the results of Kopal [9]. Both discovered the difficulty of classifying (distinguishing) the Hill and Vigenère ciphers, because of their similar statistical values.

A multi-layer classifier has been introduced by Abd and Al-Janabi [1] to classify plaintexts and ten different cipher types. The impressive results of over 99% accuracy are lessened by the enormous ciphertext length of about one million characters, which is the equivalent of an average book with 500 pages. Ciphertexts with these lengths are seldom. Based on the DECODE database, described in [14,15], the majority of encrypted historic manuscripts is only between a few lines and some pages of ciphertext long. As of the end of May 2021, the DECODE database contains more than 2,600 records of encrypted historic manuscripts and keys.

N. Krishna [10] developed approaches that have not yet been used by other authors for the four ciphers Simple Substitution, Vigenère, Transposition and Playfair. An important point for comparison is that the Hill cipher was not used here. The first approach, a support vector machine (SVM), uses the ciphertexts of length 10 to 10,000 characters, which are mapped in a number range, as training data. The SVM uses the implementation of the Sklearn Library[1] with a 10-fold-stratified-cross-validation and a one-vs-rest classifier for the calculation of the confusion matrix. This means that 9 out of 10 datasets were used for training and 1 dataset for testing in order to find the most suitable class. In the second and third approach, a Hidden Markov Model (HMM) was trained for 1,000 ciphertexts per class. In the second approach, this was used as input for a convolutional neural network, and in the third approach it was used as input for an SVM. The first approach achieves an accuracy of 100% with a text length of 200 characters, the second 71% with a text length of 155 characters and the third 100% with a text length of 155 characters.

Table 1. Summarized results and attributes of related work in the field of cipher-type detection (M = million)

Author	Accuracy in %	Ciphertext length	Dataset size	#Cipher types	Cipher category	Technology
Abd [1]	99.60	1M	N/A	11	Classical	3-level-classifier
Kopal [9]	90	100	4,500	5	Classical	FFNN
Krishna [10]	100	155	4,000	4	Classical	SVM, HMM
Leierzopf [13]	80.24	100	200M	56	Classical	FFNN, RF, NBN
Nuhn [17]	58.50	Random	1M	50	Classical	Vowpal Wabbit
Sivagurunathan [18]	84.75	1,000	900	3	Classical	FFNN

---

[1] Sklearn Library: https://scikit-learn.org/stable/.

3   Neural Cipher Identifier

During this research, the software suite "Neural Cipher Identifier (NCID)" which trains and evaluates different FE-neural-network based classifiers like FFNN, DT, RF and NBN, was extended by two FL approaches. Further details are described in [13]. The NCID software [12] is available as open-source and can be productively run for free to determine the cipher type of a given ciphertext.[2]

This section contains a short summary and the advances with respect to the previous work [13]. Basically, with the NCID software suite different types of neural networks can be trained to classify ciphertexts produced by various ciphers. The cipher types and the corresponding key lengths can be configured as well as the type of neural network to be used. The trained model gets multiple ciphertexts and predicts their cipher type. The data is generated with an extended data loader and 14 GB of English texts from the Gutenberg project[3], which can be used for research purposes free of charge. The data loader loads plaintexts with the specified length range, converts all letters to lowercase and filters all characters not included in the alphabet consisting of 26 lowercase characters. The produced ciphertext is encoded as an array of indices of the alphabet and used to calculate all features in FE approaches and directly in FL approaches. In general, neural networks need a fixed input length. The length of the ciphertext is not relevant in the FE approach, because the number of features and their lengths are always the same. In case of FL neural networks, an input is padded by repeating the ciphertext if it is not long enough. The FL neural networks should be trained with the maximal length of the expected ciphertexts, as the text easily can be extended by itself, however information is lost when the text needs to be shortened. Multiple neural network types based on Keras[4] components (FFNN) and Sklearn[5] components (DT, RF, NBN) are implemented. Also a 1-dimensional convolutional neural network has been implemented and tested, however, no notable results have been achieved with it.

3.1   Long Short-Term Memory

Hochreiter and Schmidhuber [6] describe that RNN struggle to detect long-range dependencies in sequential data, because of their short-term memory structure. Therefore, important information can be lost during processing long sequences. To overcome this problem, the long short-term memory (LSTM) was introduced [6], which uses cell states and various gates to keep or forget information. New data is filtered with the forget gate. The old hidden state together with the current input form the new hidden state by calculating several mathematical operations. The last hidden state is also the prediction of the LSTM. The forget gate uses the sigmoid activation function to decide which information should be

---

[2] https://www.cryptool.org/ncid.
[3] https://www.gutenberg.org/.
[4] Keras: https://keras.io/.
[5] Sklearn: https://scikit-learn.org/stable/.

kept and which should be forgotten. Intuitively, with the sigmoid functions values near 0 are not important, but values near 1 are kept. The input gate updates the cell state by passing the hidden state and current input into a sigmoid function. The sigmoid output then is multiplied with the output of the tanh function output of the hidden state and current input. The cell state is updated by pointwise multiplication by the forget factor and a pointwise addition with outputs of the input gate. The output gate decides what the new hidden state should be [6].

The implementation of LSTM is fully based on Keras components and uses an embedding layer with the number of classes, which is 56, as input and an output of 64 data points to create a dense representation of the ciphertext. The LSTM layer uses 500 hidden units. The output is flattened and densed into the number of classes. The softmax function finally decides on the prediction by converting inputs into a probability distribution (all positive values from 0 to 1 which add up to 1) and sorting the indices of the classes by the resulting probability.

Table 2. Iterative grid search for LSTM

Hyperparameter	Grid	Standard	Best result	Accuracy in %
Units	[50, 100, 150, 200, 500]	–	500	72.84 (68.50)
Dropout	[0, 0.2]	0	0	68.50

Table 2 shows the results of the iterative hyperparameter optimization for the LSTM neural network. The grid search showed that a higher number of units leads to better results. That's about 2% per 50 units. Due to the high computing time, however, the dropout and features were tested and compared with only 200 units, which results are shown in the brackets. It could be shown that the architecture leads to better results with a higher number of units, but the computing time increases in direct proportion to it.

3.2 Transformer

RNN process a text sequentially and therefore require relatively high computational costs to capture relationships between pieces of information located far away from each other in the sequence. The Transformer model [19] tackles this problem by using a so-called self-attention mechanism which allows to measure the influence of each word to all the other ones in the same sequence. A nice property of self-attention is that each of the scores can be computed in parallel and hence GPUs can be efficiently utilized for the training process.

There are various ways of how Transformer neural networks can be used to solve the classification problem. The actual architecture of the Transformer-based neural network is based on a Keras example [16]. This actual architecture used for cipher-type classification works like described in the following:

As the first step, all of the words are converted into 128-dimensional vectors using an embedding algorithm. Here, the vocabulary size is set to 20,000 words. The resulted vectors are passed to the self-attention layer that is used to find relations between one word vector (query) and all the other word vectors (key tensors or just keys) in the same ciphertext. Multi-headed attention with 8 heads is used, which means that 8 sets of query/keys/values weight matrices are utilized. For each of these sets, the dot product is calculated between the given query and each of the key tensors, which results in the attention scores. These scores are normalized by the softmax function to obtain attention probabilities (all positive values from 0 to 1 which add up to 1). The value vectors are multiplied by these probabilities and summed up together to produce the output of the self-attention layer for the given query. These outputs are then fed to a feed-forward neural network, with the hidden layer of size 1,024 and a ReLu activation function. This FFNN produces the output of the Transformer layer. Therefore, for each time step one vector is received. The mean across all time steps is calculated which is provided to another feed forward network on top (with the softmax activation function) for ciphertext classification.

Table 3. Iterative grid search for Transformer neural networks

Hyperparameter	Grid	Standard	Best result	Accuracy in %
Pooling	[Average, Max]	–	Average	58.73
ff hidden layer size	[1, 5]	1	1	58.73
Vocab size	[20,000, 50,000]	20,000	20,000	58.73
ff dim	[32, 64, 128, 256, 512, 1024]	32	1.024	61.96
Embed dim	[32, 64, 128, 256]	32	128	67.96
Num heads	[2, 4, 8, 16]	2	8	73.71

Table 3 shows the results of the iterative hyperparameter optimization for the Transformer neural network. Due to computational limitations the ff dim and embed dim parameters could not be further raised.

3.3   Learning-Rate Schedulers

The learning rate of a neural network optimizer can be described as the most important hyperparameter. Setting the learning rate too high means that a model is delivering poor overall performance. A learning rate that is too low means that a model has to train for a very long time before it converges. The Adam optimizer has an internal mechanism for adapting the learning rate, but it works stochastically and calculates a learning rate between 0 and the maximal value which can be adjusted [11]. Different algorithms for adapting the learning rate are discussed in this chapter. These make it possible to use a higher initial learning rate and thus significantly reduce the time required for a model to converge. There are learning-rate schedulers that start with a very small learning rate and search for the optimal learning rate in small steps in order to reduce it later on, but these are not considered due to the increase in complexity.

The time-based decay learning rate scheduler (LRS) uses a small fixed factor (decay) to adapt the learning rate after every batch (t) by multiplying it with the number of ciphertexts trained (iter). The following equation shows a calculation step after every batch [11].

$$lr_{t+1} = lr_t \cdot \frac{1}{1 + decay \cdot iter_{t+1}} \qquad (1)$$

The step-/drop-based decay LRS reduces the learning rate after a fixed number of epochs (drop), for example after every 10 epochs, by using a small decay. The following equation shows a calculation step after every epoch (t) [11].

$$lr_t = lr_0 \cdot decay^{\lfloor \frac{t}{drop} \rfloor} \qquad (2)$$

Just like with the time-based decay LRS, the exponential decay lowers the learning rate after every epoch by multiplying the initial learning rate with the exponential of a small factor (k) multiplied with the number of epochs trained (t). The main difference is that the exponential LRS reduces the learning rate rapidly in the beginning and slowly in the end. In theory that should allow the network to generalize better. The following equation shows a calculation step after every epoch [11].

$$lr_t = lr_0 \cdot e^{-kt} \qquad (3)$$

The custom step-based decay LRS uses an early stopping mechanism, which stops training as soon as no progress can be determined after 250 mini-batches. The idea is to adjust the learning rate ahead of time in order to achieve further progress. To do this, the learning rate is gradually reduced by a percentage as soon as no progress is detected after 100 mini-batches.

In NCID, the time-based decay LRS (Eq. 1) and a custom step-based decay LRS are implemented. The step-based decay LRS (Eq. 2) and the exponential LRS (Eq. 3) are not applicable in NCID, because they are based on the number of epochs trained, but training examples are generated on-the-fly and used only once, which means the epoch stays the same. The custom step-based decay LRS is used by default. It is the safest option to use, because it is only applied when no progress is made for a long time.

3.4   Transfer Learning

Transfer learning is the process of using obtained knowledge from solving one problem and transferring this knowledge to a different (but related) one. For the case of classical cipher type detection, this approach can be advantageous due to the high diversity of the historical encrypted manuscripts and ciphers which vary based on the age, language, length, presence of typos, etc. In this work, such factors are out of the scope. However, we would like to understand if the knowledge gained through our experiments can also be helpful for classifying other, but similar, ciphers which slightly differ from the ones that we used for training of the models. Transfer learning sometimes can be threatened by the so-called catastrophic forgetting, where the additional training (retraining) with a new set of classes causes known features (learned weights) to be forgotten [8]. As a first step, to determine whether catastrophic forgetting takes place in the cipher-classification problem, we evaluated whether the knowledge obtained by training the models to classify the ciphers from the smaller set, can be reused when applied to the extended sets of encryption algorithms. In order to achieve this, multiple FFNN base models were trained from scratch with 30 (B30), 35 (B35), 40 (B40) and 5 (B5) ciphers. The ciphers were selected and added alphabetically, whereas B5 only used those that were not used in B30. This results in the following scenarios:

- Extension of B30 by 5 new ciphers (T35)
- Extension of B30 by 10 new ciphers (T40)
- Extension of B30 by 5 new ciphers, however, the ciphers are added incrementally in 5 steps, whereas the newest model becomes the new basis (T35incremental)
- Training of a new model with 5 new ciphers, which are not included in B30, which is used as a basis (T5)

Table 4. Results from the transfer learning tests

Scenario	Accuracy in %	Converges after iterations in million
B30	94.44	261
B35	92.26	189
B40	85.24	166
B5	99.96	138
T35	92.28	76
T40	86.29	75
T35incremental	92.70	50–106
T5	99.98	91

Table 4 shows that transfer learning leads to substantially less training data needed for convergence and at least the same or better accuracy in all scenarios. Especially when expanding models with a large number of classes, the amount of data required drops to less than half. When comparing B35 to the scenarios T35 and T35incremental, a better accuracy and lower amount of data needed for convergence was achieved. Similarly B40 is compared to T40 and B5 is compared to T5.

3.5 Ensemble Learning

An ensemble is a model that combines the predictions of two or more models. These models can be trained in advance with the same or different datasets. The prediction can be calculated using simple procedures such as a mean voting or more complex procedures such as weighting the individual votes [4].

In NCID, mean voting and weighted voting are used as voting methods in the ensemble models. The final two ensemble models combine three FE architectures (FFNN, NB and RF) and two FL architectures (LSTM and Transformer). With weighted voting, the probabilities with the respective statistics, i.e. Precision, Recall, Accuracy, F1-Score and Matheus Correlation Coefficient, are calculated in one voting. This is intended to ensure that the architectures that are better for the respective classes are preferred. Based on statistical findings, FL architectures produce about 10–20% better results, for example, for the Myszkowski cipher, and FE architectures produce 10–20% better results, for example, for the Beaufort cipher.

Multiple models were trained with different datasets and each one was evaluated with 10 million records of the same test dataset. Table 5 shows that the FE&FL ensemble of models results in an improvement of 4% against the FE ensemble (82.78–78.67%). Compared to FFNN (the model with the best single performance), the FE-only ensemble does not improve the accuracy significantly (78.67–78.31%). However, the ensemble of FL-only approaches improves the recognition rate with about 4% (76.10–72.33%) compared to the best single FL approach, which is the Transformer neural network.

Table 5. Comparison of different models analyzing ciphertexts of 100 characters length

Model	Approach	Accuracy in %
Weighted-voting FE & FL	Ensemble	82.78
Mean-voting FE & FL	Ensemble	82.67
Weighted-voting FE-only	Ensemble	78.67
FFNN	FE	78.31
Weighted-voting FL-only	Ensemble	76.10
RF	FE	73.50
Transformer	FL	72.33
LSTM	FL	72.16
NBN	FE	52.79

4 Conclusion

The current state-of-the-art classifiers approach the cipher-type detection problem by engineering features based on domain knowledge (FE approach). In this paper, we use an FL approach exploring the two neural networks, LSTM and

Transformer, without the reliance on domain knowledge. In terms of the number and type of ciphers used, the results of this paper are mostly comparable with the classifiers by Nuhn [17] and by Leierzopf et al. [13], while the other related work use much smaller sets of ciphers or consider unrealistically long ciphertexts.

The best results for FL-only approaches were achieved with the Transformer model, which shows an accuracy of 72.33%. Applied to the same test data, this result is worse than the best FFNN FE approach [13], which achieved 78.31% accuracy. Using only a single model, FFNN outperforms the Transformer and LSTM neural networks in our particular application. However, an ensemble model, where 5 different neural network models are put together, improved the result and achieved an accuracy of 82.78%. This improvement is an indication that the FL approach detected some regularities in the data that were not explored by the FE approaches, which is an interesting result for the cryptanalysts. The improvement is also validated by the results of the ensembles of FE-only approaches with 78.67% accuracy and FL-only approaches with 76.10% accuracy. These results are summarized in Table 5.

The conclusion of this work is that the solely FE approach can be superior in fields with domain knowledge, like the classical cipher type detection. However, FL neural networks have the ability to extract unknown feature-types and can thus be a valuable additional part improving the best known FE results.

The combined FE&FL ensemble models delivered better results than any single neural network and the combination of only FE or only FL approaches. The ensemble models use the strengths of the individual models without having to spend additional computing time on training. Although the evaluation time increases significantly due to the number of models used, the combined FE&FL ensemble models lead to an improved recognition rate by another 4% which is a strong result.

The experiments with different scenarios also showed that catastrophic forgetting did not occur. This knowledge allows us to train a basic model, e.g. with all ACA ciphers, and to use it as a basis for new models. Therefore, only the last layer is exchanged with a layer of the needed output size and the pre-computed weights are further trained. As a result, we achieved the same or better accuracy with 40–60% less data and thus a much shorter training time.

Future work in this field should include the training of models with texts from different languages or texts including errors. Another interesting further research direction is to apply the FL approach to classification of modern ciphers which are designed to be resistant against the FE approach. First successes in this direction are achieved for instance by Gohr [5] against round-reduced versions of the Speck32/64 block cipher [3]. One more approach for future research is to combine different one-vs-one classifiers and add these to the decision making process.

Acknowledgements. This work has been supported by the Swedish Research Council (grant 2018–06074, DECRYPT – Decryption of historical manuscripts) and the University of Sciences Upper Austria for providing access to the Nvidia DGX-1 deep learning machine.

References

1. Abd, A., Al-Janabi, S.: Classification and identification of classical cipher type using artificial neural networks. J. Eng. Appl. Sci. **14**, 3549–3556 (2019)
2. American Cryptogram Association: Cryptogram (2005). https://www.cryptogram. org/. Visited 14 April 2021
3. Beaulieu, R., Shors, D., Smith, J., Treatman-Clark, S., Weeks, B., Wingers, L.: The SIMON and SPECK lightweight block ciphers. In: Proceedings of the 52nd Annual Design Automation Conference, pp. 1–6. Association for Computing Machinery, San Francisco California, June 2015
4. Brownlee, J.: Why use ensemble learning? October 2020. https:// machinelearningmastery.com/why-use-ensemble-learning/. Visited 14 April 2021
5. Gohr, Aron: Improving attacks on round-reduced Speck32/64 using deep learning. In: Boldyreva, Alexandra, Micciancio, Daniele (eds.) CRYPTO 2019. LNCS, vol. 11693, pp. 150–179. Springer, Cham (2019). https://doi.org/10.1007/978-3-030-26951-7_6
6. Hochreiter, S., Schmidhuber, J.: Long short-term memory. Neural Comput. **9**, 1735–1780 (1997)
7. Katz, J., Lindell, Y.: Introduction to Modern Cryptography. CRC Press, Boca Raton (2020)
8. Kemker, R., McClure, M., Abitino, A., Hayes, T., Kanan, C.: Measuring catastrophic forgetting in neural networks. 54 Lomb Memorial Drive, Rochester NY 14623, November 2017. arXiv:1708.02072
9. Kopal, N.: Of ciphers and neurons-detecting the type of ciphers using artificial neural networks. In: Proceedings of the 3rd International Conference on Historical Cryptology HistoCrypt 2020, pp. 77–86. No. 171, Linköping University Electronic Press (2020)
10. Krishna, N.: Classifying Classic Ciphers using Machine Learning. Master's thesis, San Jose State University, California, USA, May 2019
11. Lau, S.: Learning rate schedules and adaptive learning rate methods for deep learning, December 2020. https://towardsdatascience.com/learning-rate-schedules-and-adaptive-learning-rate-methods-for-deep-learning-2c8f433990d1. Visited 14 April 2021
12. Leierzopf, E.: NCID - Neural Cipher Identifier (2021). https://github.com/dITySoftware/ncid
13. Leierzopf, E., Kopal, N., Esslinger, B., Lampesberger, H., Hermann, E.: A massive machine-learning approach for classical cipher type detection using feature engineering. In: HistoCrypt (accepted) (2021)
14. Megyesi, B., Blomqvist, N., Pettersson, E.: The DECODE database: collection of historical ciphers and keys. In: In Proceedings of the 2nd International Conference on Historical Cryptology, HistoCrypt 2019, pp. 69–78. Linköping Electronic Press, Mons, Belgium, June 2019
15. Megyesi, B., et al.: Decryption of historical manuscripts: the DECRYPT project. Cryptologia pp. 1–15 (2020). https://doi.org/10.1080/01611194.2020.1716410
16. Nandan, A.: Text classification with transformer, May 2020. https://keras.io/examples/nlp/text_classification_with_transformer/
17. Nuhn, M., Knight, K.: Cipher type detection. In: Conference on Empirical Methods In Natural Language Processing, pp. 1769–1773. Doha, Quatar, October 2014

18. Sivagurunathan, G., Rajendran, V., Purusothaman, T.: Classification of substitution ciphers using neural networks. IJCSNS Int. J. Comput. Sci. Network Secur. **10**, 274–279 (2010)
19. Vaswani, A., et al.: Attention is all you need. In: 31st Conference on Neural Information Processing Systems, Long Beach, CA, USA (2017)

SOMPS-Net: Attention Based Social Graph Framework for Early Detection of Fake Health News

Prasannakumaran Dhanasekaran[1(✉)], Harish Srinivasan[2], S. Sowmiya Sree[3], I. Sri Gayathri Devi[3], Saikrishnan Sankar[1], and Vineeth Vijayaraghavan[2]

[1] SSN College of Engineering, Chennai, India
{prasannakumaran18110,saikrishnan18133}@cse.ssn.edu.in
[2] Solarillion Foundation, Chennai, India
Harish.Srinivasan@utdallas.edu,vineethv@ieee.org
[3] College of Engineering Guindy, Anna University, Chennai, India

Abstract. Fake news is fabricated information that is presented as genuine, with intention to deceive the reader. Recently, the magnitude of people relying on social media for news consumption has increased significantly. Owing to this rapid increase, the adverse effects of misinformation affect a wider audience. On account of the increased vulnerability of people to such deceptive fake news, a reliable technique to detect misinformation at its early stages is imperative. Hence, the authors propose a novel graph-based framework **SO**cial graph with **M**ulti-head attention and **P**ublisher information and news **S**tatistics **Net**work (SOMPS-Net) (https://github.com/PrasannaKumaran/SOMPS-Net-Social-graph-framework-for-fake-health-news-detection) comprising of two components – *Social Interaction Graph* (SIG) and *Publisher and News Statistics* (PNS). The posited model is experimented on the HealthStory dataset and generalizes across diverse medical topics including Cancer, Alzheimer's, Obstetrics, and Nutrition. SOMPS-Net significantly outperformed other state-of-the-art graph-based models experimented on HealthStory by 17.1%. Further, experiments on early detection demonstrated that SOMPS-Net predicted fake news articles with 79% certainty within just 8 h of its broadcast. Thus the contributions of this work lay down the foundation for capturing fake health news across multiple medical topics at its early stages.

Keywords: Fake health news · Early detection · Social network · Graph neural networks · Multi-head attention

1 Introduction

The onset of digitization has deemed social media to be a major source for news consumption. This has also resulted in the widespread diffusion of misinformation, widely known as *fake news*. [1] revealed that an agency operated dozens of

© Springer Nature Singapore Pte Ltd. 2021
Y. Xu et al. (Eds.): AusDM 2021, CCIS 1504, pp. 165–179, 2021.
https://doi.org/10.1007/978-981-16-8531-6_12

Twitter accounts masquerading as local news sources that collectively garnered more than half-a-million followers. One of the major reasons for this prevalent dissemination is stated by the *Social Identity Theory* [8]; people belonging to the same group favor the social group they are involved in. This perceived credibility hinders an individual from validating any or some of the news consumed from that group. Furthermore, the impacts of social media are not limited to politics, it also poses a huge threat to global health. A study [24] showed that 60% of US adults consumed health information from social media. The problem cascades further when misinformation is spread by credible sources.

Considering these adverse effects of misinformation, extensive research work have been carried out to tackle fake news spread. According to [2], as of 2020, majority of the existing approaches used content based models. They use the textual and visual information available in the news article. However, the methods used for fabricating fake news is evolving such that they are indistinguishable from real news. This calls for additional reliable information to be jointly explored with news content for improved fake news detection.

Social media has become the main source of news online with more than 2.4 billion internet users. Hence, Shu et al. [22] studied the user dynamics and [23] utilised the attributed information of users that are extracted from their accounts. The profile features of users such as whether the account is verified or not, number of followers and user-defined location are some of the potential factors to differentiate authentic and malicious users. Social media platforms like Twitter are used not just by millions of humans but also by innumerable bots which are designed to mimic human behaviour. According to an estimate in 2017 [18], there were 23 million bots on Twitter which comprises 8.5% of all accounts. Bots create bursts of tweets about an issue and [20] claims that they are particularly active in the early spreading phases of viral claims. They inflate the popularity of fake news and hence contribute remarkably to its spread. Also, Individual user Twitter engagements (tweets/retweets/replies) collectively help to perceive a community wide opinion about a given news article.

With increase in sensitivity of news, the social engagements associated with it increases exponentially. Such multitude of social engagements and user connections are best captured and visually represented using graphs. Graph Neural Networks (GNNs) are a class of deep learning methods designed to infer data described by graphs. GNNs have the capability to fuse heterogeneous data like engagements posted by an user and their profile activity. Hence, the inherent relational and logical information about a given news can be captured using GNN.

The main contributions of this research work are summarised as follows.

1. A novel graph-based framework that jointly utilizes social engagements of associated users along with the publisher details and social media statistics of the news article.
2. With extensive use of robust user and news features, the posited model accurately predicts the veracity of a news article at early stages to combat its spread.

3. The authors demonstrate the superiority of the proposed model over established baselines that consider health articles across various subjects.

The course of the research work is organized as follows. Section 2 discusses relevant research in fake news detection. The details of the dataset used in this work are given in Sect. 3. The problem definition is presented in Sect. 4. Section 5 elucidates the component in the proposed architecture. The results of the work and other experiments are illustrated in Sect. 6. Finally the authors conclude the work and discuss future scope of this research in Sect. 7.

2 Related Work

Fake news detection has been an active research area in the recent past. Several solutions have been proposed to detect fake news and combat its spread. Wynne et al. [28] utilized vocabulary of the news article and [3,17] used the linguistic features to detect fake news. Furthermore, the visual content present in articles has also been utilized to tackle the problem. In the real world, fake-news image differs from true-news images at both physical and semantic levels. The image quality and the characteristics of the pixel attribute to the distinction between these images at the physical and semantic level respectively. Hence, [15] captured complex patterns of fake news images in the frequency domain and extracted visual features from different semantic levels in the pixel domain for fake news detection.

Zhou et al. [30] studied the cross modal relationships between text and visual information and concluded that the two complement each other and therefore were utilized together for detecting fake news articles. However, due to limited fact-checking experts to verify a news article, improved writing style of fake news spreaders and advancements in image manipulation techniques, the aforementioned solutions are suboptimal.

An alternative approach would be to extract details about the news from a reliable platform to evaluate its authenticity. It is well established that almost every news reaches the public via social media platforms. To strengthen this claim, the fake news triangle proposed in [7], posited three items (Social network, motivation, tools & services) and claimed that without any one of these factors, fake news diffuses at a slower rate. Therefore, the diversified information available in social media provides multiperspectivity of the news to aid the detection task. Liu et al. [11] exploited user profile features from social media platforms such as Twitter and Weibo. Studies [26,29] have asserted that fake news spreads much faster than true news and also investigated the dynamic evolution of propagation topology. In support to this assertion, Wu et al. [27] used the propagation path of a news article to detect the veracity of a Twitter post.

As stated in Sect. 1, the necessity and prospects of graphs to represent data from social media has motivated several research groups [10,12,13,16] to exploit GNNs for the detection task. In [13] positive and negative knowledge graphs were constructed for detecting fake news. Further, [10] claimed that integrating user's comment and content of the article in heterogeneous graph improved the detection rate. Lu et al. [12] proposed an explainable graph model for fake news detection and Rath et al. [16] utilised trust and credibility scores of users to build a user-centric graph model for the detection task.

According to a study by Gabielkov et al. [6], 59% of the shared URLs are never clicked on Twitter and consequently, these social media users do not read the news article. Therefore, the authors of this work extensively use user profile features and social engagements (posts) information along with statistics of the news recorded in social media during the propagation of news. Social engagements such as tweets and retweets reflect the user's stance on a topic. Furthermore, the credibility of a social media user are attributed by their profile and usage statistics. The historical activity of the account can be utilized to distinguish genuine and bot accounts which further assists in detecting fake news articles. In addition to this, the meta information of news sources serves as a complementary component for fake news detection.

The authors used the HealthStory dataset to develop the proposed model that is capable of identifying health fake news articles.

3 Dataset

FakeHealth [5], is a data repository consisting of two datasets, HealthStory and HealthRelease. HealthStory contains news stories reported by news media such as Reuters Health while HealthRelease consists of official news releases from various sources such as universities, research centres and companies. The repository exclusively comprises of health related articles. HealthStory comprises of 73.6% of the news articles from FakeHealth implying that only 26.4% of the remnant articles are present in HealthRelease. Furthermore, the total number of social engagements (tweets, retweets and replies) in HealthStory is 532,380 which is significantly greater than HealthRelease which aggregates to only 65,872. Hence, authors chose to work on HealthStory for the task of fake news detection.

HealthStory includes contents of the news article, reviews about the news given by medical experts, engagements and network information for associated users. Twitter mainly consists of the following user engagements – tweet refers to an original post, retweet is a re-posting of a tweet and reply is a response to a tweet.

Medical experts rate the news based on ten independent criteria to determine the label (fake/real) of a news article. The news is rated on a 5 point scale and a rating lower than 3 implies that the news is fake. The statistics of the HealthStory dataset are provided in Table 1.

Table 1. HealthStory summary

Tweets	Retweets	Replies	Articles	True news	Fake news
384,073	120,709	27601	1,690	1,218	472

4   Problem Definition

Let $\mathcal{N} = \{n_1, n_2, n_3....n_{|\mathcal{N}|}\}$ denote the set of news articles, $\mathcal{U}$ denote the set of users who share their views on Twitter. Given a news article n_i, $\mathcal{T}^{(i)} \subseteq \mathcal{U}$ and $\mathcal{R}^{(i)} \subseteq \mathcal{U}$ be the set of users who post a tweet and retweet a tweet about the article respectively. An engagement is a tweet or retweet made by a user on the news article and is represented by $E = [\omega_1, \omega_2, \omega_3,....,\omega_e]$ where ω denotes each word in E and e is the number of words in the engagement. Let $\mathcal{P}^{(i)} = [p_1^{(i)}, p_2^{(i)}, p_3^{(i)}....p_{|f|}^{(i)}]$ denote the news (n_i) feature vector having f features. Each user $u \in \mathcal{U}$ is associated with a d-dimensional feature (e.g. user profile, historical features) vector $\mathcal{X} \in \mathbb{R}^{1 \times d}$.

Let $\mathcal{G}_t^{(i)}$, $\mathcal{G}_r^{(i)}$ be the social interaction graphs for users who engaged on the article. The subscripts denote the engagement (tweet/retweet). Each node on the graph corresponds to a user. The details of the construction of the social interaction graphs are elaborated in Sect. 5.1. $\mathcal{Y} = \{0, 1\}$ is used to denote the outcome predicted, where y = 0 indicates the news is fake and y = 1 indicates the news is real.

Given set of news article $\mathcal{N}$, $\mathcal{T}$, $\mathcal{R}$ and $\mathcal{P}$, $\mathcal{G} = \{\mathcal{G}_t, \mathcal{G}_r\}$ is obtained. This work aims to determine whether the given news article n_i is fake or real (1).

$$\mathbf{F} : \Delta(\mathcal{G}, E, \mathcal{P}) \to \widetilde{\mathcal{Y}} \tag{1}$$

This work aims to solve the problem only with the social $(\mathcal{G}, E)$ data and the data $(\mathcal{P})$ associated with the news articles.

5   Architecture

The authors propose a novel framework **SO**cial graph with **M**ulti-head attention and **P**ublisher information and news **S**tatistics **Net**work (SOMPS-Net) that consists of 2 major components; Social Interaction Graph component (SIG) and Publisher and News Statistics component (PNS) (Fig. 1).

The SIG further consists of 5 sub-components. (*i*) *Engagement Embedding* to generate a single dense vector representation for all engagements on the article. Separate vectors for tweets and retweets are obtained from this sub-component. (*ii*) *Social Connectivity Representation* to determine the social connectivity between users based on their followers and following network. (*iii*) *User Activity Matrix* contains the feature vectors of the users who engage on the article. (*iv*) *Convolutional* component to obtain the user representation in the social graph. (*v*) *Cross Attention* component to capture the correlation between the engagement and the users who engaged on the article.

The PNS is introduced to model the data associated with the news article. The truthfulness of the news (fake/real) is predicted by combining the learned representation from SIG and PNS.

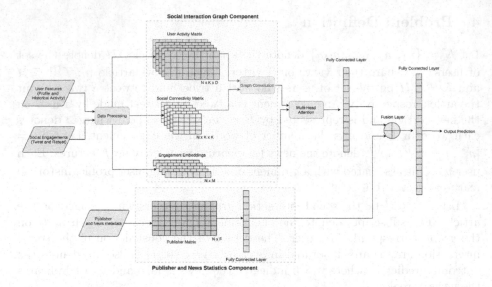

Fig. 1. SOMPS-Net

5.1 Social Interaction Graph

Engagement Embedding (EE): For every news article n_i there are p tweets and q retweets posted on Twitter and they have a maximum length of 280 characters. Since each engagement has a variable number of characters, the authors propose a method to handle the inconsistency. Firstly the content of the tweet is tokenized and padded to a maximum sequence length of m which is determined by the median number of words used in all the tweets on the article. The GloVe model [14] is then used to generate the vector representation of each element in the resulting sequence. The embeddings of all the tweets are then combined and the resultant is a 3-dimensional vector $_tV^{(i)} \in \mathbb{R}^{p \times m \times e}$, where e is the embedding dimension. Finally, a dense vector $_t\bar{V}^{(i)} \in \mathbb{R}^{m \times e}$ is obtained by averaging the embedding values across every tweet posted about the article given by Equation (2). Similarly, the vector $_r\bar{V}^{(i)}$ is obtained for the retweets made on the article. The averaged engagement embeddings ($\bar{\mathcal{V}}$) obtained from (2) are then passed through a Bi-Directional LSTM [19] layer. Let $\mathcal{L}$ denote the output from the Bi-LSTM layer.

$$_t\bar{V}^{(i)} = \frac{\Sigma_{j=1}^p {_tV_j^{(i)}}}{p},$$

(2)

where $_tV_j^{(i)}$ is the word embedded vector representation of a tweet j.

Social Connectivity Representation (SC): For each news n_i let $\tau_t^{(i)}$ and $\tau_r^{(i)}$ be the timestamp at which the first tweet and retweet were made about the article. The first k_t users who tweet about the article since $\tau_t^{(i)}$ are considered for the experiments. The number of users considered is determined by obtaining the median value given by the formula $k_t = \text{med}\{|\ \mathcal{T}^{(1)}\ |, |\ \mathcal{T}^{(2)}\ |...|\ \mathcal{T}^{(|\mathcal{N}|)}\ |\}$.

Consider two users u_x and u_y from the first k_t users. Let $\phi(u_x)$ denote the set of users who follow u_x and $\psi(u_x)$ denote the set of users followed by u_x. The connectivity score between u_x and u_y is given by Eq. (3). As a result, social connectivity matrix $\mathcal{A}_t^{(i)} \in \mathbb{R}^{k_t \times k_t}$ is obtained for the users who tweeted about the news article. Similarly, the social connectivity matrices $(\mathcal{A}_r)$ for retweet users are obtained.

$$Connect(u_x, u_y) = \frac{|\ (\phi(u_x) \cap \phi(u_y)) \cup (\psi(u_x) \cap \psi(u_y))\ |}{|\ \phi(u_x) \cup \phi(u_y) \cup \psi(u_x) \cup \psi(u_y)\ |} \qquad (3)$$

User Activity Matrix (UAM): Each user who posts about the article is associated with a feature vector $(\mathcal{X})$ as mentioned in Sect. 4. In addition to the available profile features, the authors extract features based on historical activity of the account. Features considered are listed in Table 2. As a result, node feature matrices $\mathcal{H}_t^{(i)} \in \mathbb{R}^{k_t \times d}$ and $\mathcal{H}_r^{(i)} \in \mathbb{R}^{k_r \times d}$ are generated to represent the node (user) features for a news article (n_i).

Table 2. UAM features

Whether the account is protected or not	Whether the profile image is default or not
Whether the account is verified or not	Whether the account UI is default or not
Whether geo location is enabled or not	Number of words in the description of the user
Number of words in the username	Number of tweets liked by the user
Friends count of the user	Average number of posts made per day
Followers count of the user	Maximum number of posts on a single day
No. public lists the user is member of	Days between account creation and post
Number of tweets made on the news article (TW only)	Time between first tweet and retweet in hours (RT only)

Graph Convolution (GC): For each n_i, graph $\mathcal{G}^{(i)}$ is defined by its connectivity matrix $\mathcal{A}^{(i)}$. Two independent graphs $(\mathcal{G}_t, \mathcal{G}_r)$ for each engagement (tweet/retweet) are obtained. A graph is $\mathcal{G}$ is represented by a tuple containing set of nodes/vertices and edges/links. The graph can be represented as $\mathcal{G} = (\mathcal{V}, \mathcal{E})$. The nodes of the graph $(\mathcal{V})$ represent the users and the edges $(\mathcal{E})$ represent the connectivity score (3) between the users. A degree normalized social connectivity matrix $(\bar{A})$ is derived using Eq. (4) from the degree matrix $\mathcal{D}$. A Graph Convolutional Network (GCN) layer [9] is then applied over $\mathcal{G}$ to obtain the graph embeddings. The number of layers in a GCN corresponds to the farthest distance that the node features can propagate. For this work, the authors consider using a 3 layered stacked GCN to capture the finer representation of nodes in the graph network. Let l denote the lth layer of the stacked GCN. The node representations in the $(l + 1)^{th}$ layer is given by Eq. (5). Consequently, the graph representations $\mathcal{G}_t$, $\mathcal{G}_r$ corresponding to $\mathcal{A}_t$, $\mathcal{A}_r$ are obtained.

$$\bar{A} = \mathcal{D}^{\frac{-1}{2}} \mathcal{A} \mathcal{D}^{\frac{-1}{2}}, \tag{4}$$
$$\mathcal{H}^{(l+1)} = \sigma(\bar{A}\mathcal{H}^{(l)}\mathcal{W}^{(l)}) \tag{5}$$

where,

$W^{(l)}$ is the weight matrix of the ith graph convolutional layer
$\mathcal{D}_{ii} = \Sigma_j \mathcal{A}_{ij}$
σ is the activation function

Multi Head Attention (MHA): Attention mechanism can be described as the weighted average of (sequence) elements with weights dynamically computed based on an input query and element's key. Query (Q) corresponds to the sequence for which attention is paid. Key (K) is the vector used to identify the elements that require more attention based on Q. The attention weights are averaged to obtain the value vector (V). A score function (6) is used to determine the elements which require more attention. The score function takes Q and K as input and outputs the attention weight of the query-key pair. In this work, the authors consider using the scaled dot product proposed by Vaswani et al. [25]. The attention weights are calculated based on the graph embeddings ($\mathcal{G}$) and the Bi-LSTM output ($\mathcal{L}$). K and V are initialized with the value of $\mathcal{G}$ and Q is initialized with the value of $\mathcal{L}$.

The scaled dot product attention captures the characteristics of the sequence it attends. However, often there are multiple different aspects to a sequence, and these characteristics cannot be captured by a single weighted average vector. The Multi-Head Attention (MHA) [25] mechanism resolves this by performing the attention function (6) in parallel on multiple heads. This allows the model to jointly attend to information from different representation sub-spaces at different positions. The queries, keys and values are linearly projected into sub-queries, sub-keys and sub-value pairs and different learned representations are obtained from each head (7). The attention output obtained from each head is then combined and the final weight matrix (W^O) is calculated. d_k represents the hidden dimensionality of K. Thus the output O^S from the SIG component is obtained.

$$Attention(Q, K, V) = Softmax(\frac{QK^T}{\sqrt{d_k}})V \tag{6}$$

$$MultiHead(Q, K, V) = Concat(head_1, \ldots, head_n)W^O$$
$$\text{where } head_i = Attention(QW_i^Q, KW_i^K, VW_i^V), \tag{7}$$
$$W^Q, W^K, W^V \text{ are the weight matrices of Q, K and V respectively}$$

5.2 Publisher and News Statistics

The Publisher and News Statistics (PNS) is the second component in the SOMPS-Net framework. The intention of the fake news spreader is to proliferate the news instantaneously and provoke chaos amongst the targeted audience. Thus, the authors propose that utilizing the statistical information of the news, recorded throughout its lifetime along with the metadata of the news article could help in the detection task. Further, credibility information about a news publisher is another significant factor that helps in determining the authenticity of the news. The features considered are listed in Table 3. The feature vector $(\mathcal{P})$ is passed through a dense layer and the output O^P from this component is obtained.

Table 3. PNS features

Total number of tweets	Number of unique users mentioned
Total number of retweets	Lifetime of the news in days
Total number of replies	Tags associated with article
Total number of unique hashtags	News publisher
Total number of likes (any engagement)	Average rating of the news publisher

The outputs O^S and O^P are fused together and the resulting high dimensional vector is passed through a fully connected layer. The final outcome $\mathcal{Y}^{(i)}$ for n_i is predicted by the model as illustrated in Eq. (8).

$$\Gamma(O^S \oplus O^P) \rightarrow Y \tag{8}$$

6   Experiments and Results

6.1   Experimental Setup

The performance of SOMPS-Net is evaluated and compared based on Accuracy and F1-score. The news articles were proportionally sampled (stratified) and the data was split into 75% for training, 10% for validation and 15% for testing. The authors consider news articles that have at least one of each engagement (tweet/retweet). As a result, 1492 news articles $(\mathcal{N})$ were obtained. The number of real and fake news articles obtained were 1082 and 410 respectively. The model configuration is listed in Table 4.

6.2   Comparison Systems

We compare the performance of the proposed architecture with the previous works done on HealthStory.

Table 4. Model configuration

Number of tweet users (k_t) : 118	Number of retweet users (k_r) : 12
Engagement (tweet/retweet) length (m) : 20	Number of news features (f) : 10
Number of user associated features (d) : 15 (tweet)	Number of user associated features (d) : 15 (retweet)
Word embedding dimension (e) : 100	Number of GCN layers (l) : 3
GCN output dimension : 16	Number of hidden units in Bi-LSTM : 100
Number of attention heads (n) : 16	Size of each attention head for key (K) and query (Q) : 4
Size of each attention head for query (Q) : 4	Size of attention head for value (V) : 12
Learning rate : 0.005	Dropout : 0.5
Optimizer : SGD (Gradient descent with momentum)	Loss function : Binary cross entropy

Dai et al. [5] considered three methods – (i) linguistic-based, (ii) content-based and (iii) social context-based for fake news detection. In linguistic based methods, the authors used Logistic Regression, SVM and Random Forest for modelling the lexicon-level features. CNN and Bi-directional GRU were employed for content-based modelling. Finally, in social context based methods, the authors used the Social Article Fusion (SAF) model initially proposed by [21]. The SAF model utilizes user embeddings and replies. The social context features learned from an LSTM encoder were combined with the former to make the final prediction.

Chandra et al. [4] proposed a framework SAFER which uses graph-based model for fake news detection. The framework aggregates information from the content of the article, content sharing behaviour of users and the social connections of the users. SAFER consists of two components – graph encoder and text encoder component. The graph encoder takes the community graph of the users and the text encoder takes the text of the article as inputs. The outputs from the two components are then concatenated and passed through a logistic classifier. The authors considered six different GNN architectures for generating user embeddings.

SOMPS-Net (This Work). The results using the proposed novel framework SOMPS-Net are compared with other systems for its robustness. To validate the importance of each of the components in the proposed framework, the authors also consider 2 variants of SOMPS-Net – $SOMPS_{SIG}$ and $SOMPS_{PNS}$. In $SOMPS_{SIG}$ only the SIG component is considered and in $SOMPS_{PNS}$ only the PNS component is considered. Equation (9) represents the components and the data used in the 3 variants.

$$F_{SOMPS} : \Delta(\mathcal{G}, E, \mathcal{P}) \to \tilde{y}$$
$$F_{SOMPS_{SIG}} : \Delta(\mathcal{G}, E) \to \tilde{y} \qquad\qquad (9)$$
$$F_{SOMPS_{PNS}} : \Delta(\mathcal{P}) \to \tilde{y}$$

6.3 Results Analysis

Table 5 illustrates the results obtained using SOMPS-Net and other comparison systems considered.

The SOMPS-Net framework outperforms linguistic-based and content-based models used in Dai et al. [5] by 6.1% and 6.6% respectively. This further solidifies the initial hypothesis of considering the social engagement data to model fake news detection. On comparison, social context-based model used in [5] has around 4% accuracy improvement over SOMPS-Net. However, SOMPS-Net performs better than this model by 4% when F1 score is considered. Moreover, the social context model uses replies made on the article. Upon exploratory analysis it was found that only 720 articles contained replies. Thus an accuracy of 76% was achieved using only 541 true and 179 fake news articles. On the other hand, the authors of this work considered 1492 articles as mentioned in Sect. 6.2. Since, the social context model considered by [21] uses less than 50% of the HealthStory articles, it has limited applicability.

SAFER achieved an F1-score of 62.5%, and SOMPS-Net outperformed SAFER with a relative 27.36% performance improvement, asserting its superior performance. Also, it can be inferred from Table 5 that SOMPS-Net outperforms each of the six different GNN architectures considered by SAFER. Furthermore, the robustness of the SOMPS-Net's components are well established since each of the components in SOMPS-Net – $SOMPS_{SIG}$ and $SOMPS_{PNS}$ outperformed SAFER by 16% and 12.2% respectively. Additionally, it is also observed that SAFER uses the news content for modelling while SOMPS-Net uses only the social context and metadata information about the news. This proves the initial hypothesis of considering only the social context and publisher information for detecting fake news articles since SOMPS-Net significantly outperformed SAFER.

Table 5. Fake news detection results

Model	Approach	Accuracy	F1 Score
Dai et al.	Linguistic-based	0.720	0.735
Dai et al.	Content-based	0.742	0.730
*Dai et al.	Social context-based	0.760	0.756
SAFER	Graph + Content	Not specified	0.625
SOMPS- P	Publisher & News	0.727	0.747
SOMPS- SIG	Graph	0.727	0.785
SOMPS-Net	**Graph + Publisher & News**	**0.727**	**0.796**

* Uses only 720 (42.6%) articles

6.4   Early Detection

Early detection of fake news is crucial to restrain its reach from wider audience, particularly for health related information. The task of early detection is driven by social engagements of the news article captured within a fixed time frame. Time intervals in multiples of 4 since the first tweet about the article were considered. The performance of SOMPS-Net framework for each time interval is illustrated in Fig. 2.

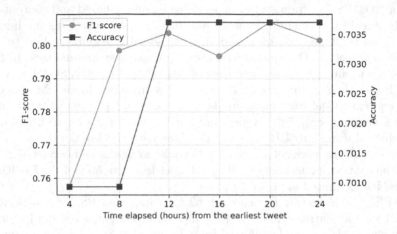

Fig. 2. Early detection results

From Fig. 2 the authors infer the following. The detection rate of the model in terms of F1-score improves drastically till 8 h and sustains around 80%. Maximum F1-Score of 0.807 was achieved from 20 h of its broadcast. It can be noted that the model detects with an appreciable score of 75% within just 4 h which is higher than the comparison systems mentioned in Sect. 6.2. Hence, it is evident that SOMPS-Net is robust and is capable of detecting fake health articles at its early stages with minimal information. This analysis could therefore help control further diffusion of fake news in social media.

7   Conclusion

In this work, the authors propose and successfully test a novel graph based framework – SOMPS-Net on the HealthStory dataset to detect fake news in the health domain. The proposed framework utilizes the social context data and social reach data of the news article. It consists of 2 components – Social Interaction Graph (SIG) which consists of 5 sub-components (Sect. 5.1) utilizing the social context data such as tweets, retweets and user profile features. The authors use a rich user profile feature set (as illustrated in Table 2) that contains user metadata and features extracted from the historical activity of the account. The

Publisher and News Statistics (PNS) component (Sect. 5.2) utilizes the metadata and statistics of the news article (illustrated in Table 3).

SOMPS-Net performs significantly better than other well-established graph based approaches on HealthStory. A 27.36% relative performance improvement is achieved from the state-of-the-art graph based models. The importance of each of the components is established since the components – SIG and PNS outperform SAFER by 16% and 12.2% respectively. The authors further exhibit SOMPS-Net's robustness by using data captured within a certain time frame. The model detected fake news with 79.8% certainty with only 8 h of data and achieved a maximum F1-score of 80.7% with 20 h of information.

For future work, the authors aim to include other modalities into the proposed framework. Also, there exists a need for interpretable and explainable machine learning solutions for health news articles. Providing deep insights about the model and an explanation on why the news is classified as fake or real can be critical in mitigating fake news spread and also safeguard the consumers from adversaries.

References

1. Badawy, A., Ferrara, E., Lerman, K.: Analyzing the Digital Traces of Political Manipulation: The 2016 Russian Interference Twitter Campaign, p. 258–265. IEEE Press (2018)
2. Dylan de Beer, M.M.: Approaches to identify fake news: a systematic literature review. ACM Trans. Comput. Syst. **32**(2) (2020). https://doi.org/10.1007/978-3-030-49264-9_2
3. Bhutani, B., Rastogi, N., Sehgal, P., Purwar, A.: Fake news detection using sentiment analysis. In: 2019 Twelfth International Conference on Contemporary Computing (IC3), pp. 1–5 (2019). https://doi.org/10.1109/IC3.2019.8844880
4. Chandra, S., Mishra, P., Yannakoudakis, H., Nimishakavi, M., Saeidi, M., Shutova, E.: Graph-based modeling of online communities for fake news detection. CoRR abs/2008.06274 (2020). https://arxiv.org/abs/2008.06274
5. Dai, E., Sun, Y., Wang, S.: Ginger cannot cure cancer: battling fake health news with a comprehensive data repository. arXiv preprint arXiv:2002.00837 (2020)
6. Gabielkov, M., Ramachandran, A., Chaintreau, A., Legout, A.: Social clicks: what and who gets read on twitter? ACM SIGMETRICS Perform. Eval. Rev. **44**, 179–192 (2016). https://doi.org/10.1145/2964791.2901462
7. Gu, L., Kropotov, V., Yarochkin, F.: The fake news machine: how propagandists abuse the internet and manipulate the public. Trend Micro. **5**, 1–85 (2017)
8. Hogg, M.A.: Chapter 5 Social Identity Theory: pp. 112–138. Stanford University Press (2020). https://doi.org/10.1515/9781503605626-007
9. Kipf, T.N., Welling, M.: Semi-supervised classification with graph convolutional networks. In: International Conference on Learning Representations (ICLR) (2017)
10. Liao, H., Liu, Q., Shu, K., Xie, X.: Incorporating user-comment graph for fake news detection. CoRR abs/2011.01579 (2020). https://arxiv.org/abs/2011.01579
11. Liu, Y., Wu, Y.F.B.: FNED: a deep network for fake news early detection on social media. ACM Trans. Inf. Syst. **38**(3) (2020). https://doi.org/10.1145/3386253
12. Lu, Y.J., Li, C.T.: GCAN: graph-aware co-attention networks for explainable fake news detection on social media (2020)

13. Pan, J.Z., Pavlova, S., Li, C., Li, N., Li, Y., Liu, J.: Content based fake news detection using knowledge graphs. In: Vrandečić, D., et al. (eds.) ISWC 2018. LNCS, vol. 11136, pp. 669–683. Springer, Cham (2018). https://doi.org/10.1007/978-3-030-00671-6_39

14. Pennington, J., Socher, R., Manning, C.D.: GloVe: global vectors for word representation. In: Empirical Methods in Natural Language Processing (EMNLP), pp. 1532–1543 (2014). http://www.aclweb.org/anthology/D14-1162

15. Qi, P., Cao, J., Yang, T., Guo, J., Li, J.: Exploiting multi-domain visual information for fake news detection. In: 2019 IEEE International Conference on Data Mining (ICDM), pp. 518–527 (2019). https://doi.org/10.1109/ICDM.2019.00062

16. Rath, B., Morales, X., Srivastava, J.: SCARLET: explainable attention based graph neural network for fake news spreader prediction. In: PAKDD (2021)

17. Rubin, V., Conroy, N., Chen, Y., Cornwell, S.: Fake news or truth? Using satirical cues to detect potentially misleading news. In: Proceedings of the Second Workshop on Computational Approaches to Deception Detection, pp. 7–17. Association for Computational Linguistics, San Diego, California, June 2016. https://doi.org/10.18653/v1/W16-0802, https://aclanthology.org/W16-0802

18. Salge, C.: Is that social bot behaving unethically? Commun. ACM **60**, 29–31 (2017). https://doi.org/10.1145/3126492

19. Schuster, M., Paliwal, K.: Bidirectional recurrent neural networks. IEEE Trans. Sign. Process. **45**(11), 2673–2681 (1997). https://doi.org/10.1109/78.650093

20. Shao, C., Ciampaglia, G., Varol, O., Flammini, A., Menczer, F., Yang, K.C.: The spread of low-credibility content by social bots. Nat. Commun. 9 (2018). https://doi.org/10.1038/s41467-018-06930-7

21. Shu, K., Mahudeswaran, D., Liu, H.: FakeNewsTracker: a tool for fake news collection, detection, and visualization. Comput. Math. Organ. Theory **25**, 60–71 (2019)

22. Shu, K., Wang, S., Liu, H.: Beyond news contents: the role of social context for fake news detection. In: Proceedings of the Twelfth ACM International Conference on Web Search and Data Mining, pp. 312–320. WSDM 2019. Association for Computing Machinery, New York, NY, USA (2019). https://doi.org/10.1145/3289600.3290994

23. Shu, K., Zhou, X., Wang, S., Zafarani, R., Liu, H.: The role of user profiles for fake news detection. In: Proceedings of the 2019 IEEE/ACM International Conference on Advances in Social Networks Analysis and Mining, pp. 436–439. ASONAM 2019. Association for Computing Machinery, New York, NY, USA (2019). https://doi.org/10.1145/3341161.3342927

24. Thackeray, R., Crookston, B., West, J.: Correlates of health-related social media use among adults. J. Med. Internet Res. **15**, e21 (2013). https://doi.org/10.2196/jmir.2297

25. Vaswani, A., et al.: Attention is all you need. In: Advances in Neural Information Processing Systems, pp. 5998–6008 (2017)

26. Vosoughi, S., Roy, D., Aral, S.: The spread of true and false news online. Science **359**(6380), 1146–1151 (2018). https://doi.org/10.1126/science.aap9559, https://science.sciencemag.org/content/359/6380/1146

27. Wu, L., Liu, H.: Tracing fake-news footprints: characterizing social media messages by how they propagate. In: Proceedings of the Eleventh ACM International Conference on Web Search and Data Mining, pp. 637–645. WSDM 2018. Association for Computing Machinery, New York, NY, USA (2018). https://doi.org/10.1145/3159652.3159677

28. Wynne, H.E., Wint, Z.Z.: Content based fake news detection using n-gram models. In: Proceedings of the 21st International Conference on Information Integration and Web-Based Applications & amp; Services, pp. 669–673. iiWAS2019. Association for Computing Machinery, New York, NY, USA (2019). https://doi.org/10.1145/3366030.3366116

29. Zhao, Z., et al.: Fake news propagate differently from real news even at early stages of spreading. EPJ Data Sci. **9** (2018). https://doi.org/10.1140/epjds/s13688-020-00224-z

30. Zhou, X., Wu, J., Zafarani, R.: SAFE: similarity-aware multi-modal fake news detection. In: Lauw, H.W., Wong, R.C.-W., Ntoulas, A., Lim, E.-P., Ng, S.-K., Pan, S.J. (eds.) PAKDD 2020. LNCS (LNAI), vol. 12085, pp. 354–367. Springer, Cham (2020). https://doi.org/10.1007/978-3-030-47436-2_27

The Impact of Sentiment in the News Media on Daily and Monthly Stock Market Returns

Justin Case[(✉)] and Adam Clements

School of Economics and Finance, Queensland University of Technology,
2 George Street, Brisbane City, QLD 4000, Australia
{jj.case,a.clements}@qut.edu.au

Abstract. Sentiment analysis allows for the subjective information contained within a piece of media to be classified. This subjective information is particularly relevant in the finance domain as opinions and speculation may influence investment decisions and, consequently, affect asset prices. State-of-the-art sentiment classification methods in machine learning use contextual word representations from pre-trained BERT language models and several papers have fine-tuned BERT for the sentiment classification of financial texts. However, these preceding studies have not considered extensive real-world data sets or provided a robust assessment of whether the sentiment scores derived from BERT predict asset prices. This paper addresses these limitations by fine-tuning BERT to analyse the sentiment across an extensive set of news articles published by the Wall Street Journal. The analysis also extends beyond financial news stories, assessing the sentiment in economic and national news articles. An econometric evaluation of the sentiment scores shows that financial news sentiment is statistically significant in predicting daily market returns, while longer-term economic news sentiment predicts monthly returns.

Keywords: Sentiment analysis · Asset pricing · BERT

1 Introduction

Within financial economics there is a significant body of literature which explores the impact of media sentiment on asset prices[1]. Shiller [39] argues that this relationship exists because investors follow the narratives contained within the media, even though these narratives may be pure speculation. The notion that sentiment influences asset prices, however, contradicts neoclassical asset pricing theory, which argues that asset prices reflect fundamentals and that investors are informed and rational [14,15].

A central challenge in assessing the relationship between media sentiment and asset prices is to accurately identify the subjective sentiment. For the automated analyses of textual data machine learning (ML) approaches offer a significant theoretical advantage over lexicon approaches. This is because ML approaches

---

[1] For reviews of textual analysis in financial economics see [1,19,30,44].

© Springer Nature Singapore Pte Ltd. 2021
Y. Xu et al. (Eds.): AusDM 2021, CCIS 1504, pp. 180–195, 2021.
https://doi.org/10.1007/978-981-16-8531-6_13

are able to capture the contextual characteristics of an expression that contribute to the sentiment, whereas lexicon approaches cannot. State-of-the-art ML approaches for sentiment classification use contextual word representations from pre-trained BERT (Bidirectional Encoder Representations from Transformers) language models [11].

Several previous papers have fine-tuned BERT for the sentiment classification of financial texts. However, these studies have not utilised extensive real-world data sets or provided a robust analysis of the relationship between the sentiment scores estimated with BERT and asset prices. More generally within the literature there has been limited research on how the sentiment within different news topics impact asset prices or how short-term and longer-term sentiment trends affect short-run and long-run asset prices, respectively.

This paper proposes to address the limitations of previous studies. The media corpus analysed with BERT in this paper is comprised of Wall Street Journal (WSJ) news articles published between January 1, 2015 and January 1, 2021. Furthermore, to gain a detailed understanding of the relationship between media sentiment and asset prices the sentiment is analysed across financial, economic, and national news topics. Different frequency components of both sentiment and stock returns are also explored in the analysis, which is conducted within an econometric framework.

The results presented in this paper illustrate that the sentiment in financial news items published in the six-hours before trading is predictive of daily price changes in the S&P 500 index. Furthermore, there is a statistically significant negative relationship between long-term economic sentiment and long-run stock prices. We speculate that this negative relationship may be due to economic downturns leading to fiscal stimulus and lower interest rates, which in turn push up stock prices. The results presented in this paper therefore begin to provide a more detailed understanding of the complex relationship between the news media and asset prices.

2   Related Literature

2.1   Feedback from Text Media to Financial Markets

There is an established body of literature on the textual analysis of media content to predict stock market movements. The majority of these studies have focused on the content of newspaper articles due to the news media's central role in disseminating information and the availability of historical news articles. However, the analysis of alternative media content is becoming more common, particularly online and social media, as the significance of these mediums increase.

Tetlock [43] presents one of the earliest studies on automated textual analysis to determine the sentiment in news articles. To classify the sentiment within each article Tetlock utilises the Harvard IV-4 Psychosocial Dictionary, which is narrowed from 77 predefined categories to a focus on a composite category of words with a negative outlook. Tetlock finds that from 1984 through 1999 negative words in the 'Abreast of the Market' column in the WSJ are associated with lower same-day stock returns and predict lower returns the following day.

Garcia [17] expands on Tetlock's [43] study by analysing the sentiment in two New York Times financial columns over the period from 1905 to 2005. Garcia employs the Loughran-McDonald dictionary [29], which accounts for the nuances of financial jargon, to classify the sentiment in each article. As with Tetlock [43], Garcia [17] finds that linguistic tone predicts the following day's return on the stock market. Furthermore, Garcia shows that the predictability of stock returns using news content is concentrated during recessions, supporting experimental research from psychology that anxiety increases an individual's likelihood to seek and rely on advice [20]. Studies similar to those in Tectlock [43] and Garcia [17] can also be found in [12,21,41].

A significant drawback of the studies by Tetlock [43] and Garcia [17] is the exclusive focus on daily sentiment contained within one or two columns of a newspaper. Larsen and Thorsrud [28] present a broader study by decomposing the articles in a Norwegian newspaper into latent topics and then investigating the significance of each topic in explaining economic fluctuations. They find that the sentiment within several latent topics is predictive of key economic variables, including asset prices. The study, however, is based on a general lexicon, which discounts any domain specific language when assessing the sentiment. This challenge of optimising sentiment quantification for a given domain has been the focus of much of the recent literature, as further discussed in Sect. 2.2.

In addition to traditional news media, there have been an increasing number of studies analysing the sentiment contained in alternative media sources. This research follows the work of Antweiler and Frank [2] and Das and Chen [10], which show the predictability of stock markets using data from Internet message boards. Since these studies there have been mixed findings on the predictability of asset prices using online information, as summarised in [33]. A significant issue that faces research utilising online or social media data is noise in the data set (see [8,31] for further details).

2.2   Computational Methods for Sentiment Analysis

The two main approaches to sentiment classification in finance and economic studies are lexicon and ML approaches. Lexicon-based approaches are defined by words (or sequences of words) with an associated sentiment score. The advance in these approaches has been the shift from general lexicons to domain-specific lexicons in an effort to optimise sentiment quantification. Loughran and McDonald [29] illustrate the limitations of utilising a general lexicon by showing that the Harvard Psychosociological Dictionary, as employed by Tetlock [43] and Larsen and Thorsrud [28], misclassifies a significant number of words commonly used in a financial context.

Within the recent finance and economics literature there have been several statistical methodologies explored to enhance domain specific lexicons. For example, Jegadeesh and Wu [24] infer word importance from market reactions, and Shapiro et al. [38] and Renault [36] use statistical methods for automatic domain specific lexicon generation. Lexicon methods are in general, however, unable to determine the linguistic context around which the sentiment words appear. These methods therefore fail analyse the deeper semantic meaning of a given text.

ML sentiment classification approaches are theoretically able to overcome the shortfalls of lexicon-based methods. This is because ML approaches are able to learn the sentiment weights for words and phrases, and how these weights combine to define the sentiment of a complete expression. As with lexicon-based methods, ML models are ideally adapted for specific domains in order to optimise sentiment quantification. For supervised learning models this requires a separate annotated data set for each domain.

There are numerous ML approaches to sentiment classification. Supervised models that have been employed include Naïve Bayes, maximum entropy classification, and support vector machines (SVMs). Recently, neural networks, and in particular deep learning algorithms, have become more prominent. A detailed survey of ML-based methods for the analysis of textual data, with a focus on economic research, is given in [19].

One particular deep learning algorithm that has achieved state-of-the-art results across numerous natural language processing (NLP) tasks is BERT [11]. BERT is a deep neural network that is able to explicitly integrate global and local context into word embeddings. The model has previously been fine-tuned for sentiment classification in the finance domain [3, 42, 46] and has been shown to outperform other ML models. However, the application of BERT to an extensive real-world data set in order to examine the relationship between news media sentiment and asset prices has yet to be performed.

3   Methodology

The following sections detail the methodology we have employed to gain a deeper understanding of how sentiment in the news media influences asset prices. The following sections cover: the data set used; how the relevant news topics and their characteristics have been identified; and the fine-tuning of BERT for sentiment classification across the chosen news topics. The econometric methods used to interpret the sentiment scores determined by BERT are also discussed.

3.1   The Data

This study has used two principal sources of data. The first is stock price data from the S&P 500 index that covers the analysis period from January 1, 2015 to January 1, 2021. The stock price data, namely the daily opening and closing prices of the S&P 500, was collected from Yahoo Finance. From this data the daily open-to-close returns on the index r_t^d were calculated as the log difference between the opening price and the closing price

$$r_t^d = log(p_{c,t}) - log(p_{o,t}) \tag{1}$$

where $p_{c,t}$ and $p_{o,t}$ are the daily closing and opening prices, respectively. The monthly returns on the index were calculated as

$$r_t^m = log(p_{m,t}) - log(p_{m,t-1}) \tag{2}$$

where r_t^m is the monthly return and $p_{m,t}$ is the closing price of the index on the final day of the month. The second data source relates to the news media content analysed in this paper.

The news media used in this study was taken from the digital WSJ achieve, which goes back as far as 1997. The data is available to any subscriber of the WSJ, as well as through other media providers such as ProQuest and Dow Jones. The data set was built by downloading the content of the WSJ articles into a database. The attributes recorded of each article include: the date originally published; the date updated; the authors; the title; the summary text; the full text of the article; the newspaper section under which the article was published; and the keywords associated with the article. This study has analysed the 235,040 articles published by the WSJ between January 1, 2015 and January 1, 2021.

3.2   Topic Modelling of the News Media Corpus

To identify the topics in the WSJ corpus a Latent Dirichlet Allocation (LDA) model has been employed. This model, which is commonly applied in the ML literature, was selected for its simplicity and effectiveness. LDA [4] is an unsupervised topic model that clusters words into topics and at the same time classifies each article in the corpus as a mixture of topics. Employing the statistical semantics package Gensim [35] to implement the LDA model, the WSJ corpus was classified into 13 latent topics[2]. We selected 13 topics as this approximately aligns with the number of news sections into which the WSJ classify their articles over the analysis period.

Table 1 lists the 13 latent topics from the LDA analysis together with the most relevant words in each topic. It should be noted that the topic labels stated in Table 1 are subjective as the LDA estimation procedure does not label the latent topics. The labels assigned to each topic were based on an interpretation of the topic's word cluster. A graphical depiction of the similarity between the topics, as measured by the Jensen-Shannon divergence, is illustrated in Fig. 1.

As discussed in Sect. 2, both lexicon and ML techniques are ideally adapted for specific domains in order to optimise sentiment quantification. This requires a separate annotated data set for each domain. Utilising the results from the LDA we therefore narrowed our study and selected three news topics likely to impact the price of the S&P 500 index–namely financial markets (Topic 6), economic activity (Topic 5), and US national issues (Topics 8, 11, and 12). For these three news topics a domain-specific classifier has been employed to optimise the sentiment quantification of the WSJ news articles, as further discussed in Sect. 3.3.

The selection of the three news topics analysed in this study is supported by previous literature. The significance of feedback from financial news to the

---

[2] The LDA analysis was performed with a batchsize of 1000, an asymmetric alpha hyperparameter, and 10 passes over the corpus. These parameters were selected as they generate the most interpretable word clusters. The pre-processing steps taken for the LDA included; tokenization; the removal of stopwords; and lemmetization.

Table 1. Latent topics in the WSJ.

ID	Label	Relevant words
1	Politics	leader, political, election, campaign, vote, party
2	Interest	game, season, art, life, team, women, book
3	Business	deal, firm, company, investment, business, fund
4	Legal	tax, Worker, system, employee, law, bill
5	Economic Activity	economy, job, growth, trade, economic, activity
6	Markets	profit, stock market, share, revenue, loss, investor
7	Monetary Policy	oil, dollar, interest rate, oil price, inflation, gold
8	Order	case, police, authority, death, court, investigation
9	World News	bank, debt, government, military, migrant, region, crisis
10	Lifestyle	home, food, property, store, brand, retailer
11	Innovation	drug, car, startup, treatment, patient, research
12	Environment	regulator, standard, insurance, emission, coal, land
13	Credit	bankruptcy, creditor, settlement, initial public offering

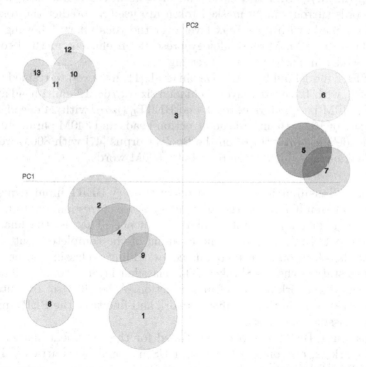

Fig. 1. The LDA topics identified in Table 1 visualised with LDAvis [40]. The scaling of the inter-topic distances is defined by principal components (PC). The topics selected for sentiment classification in this study are highlighted: Topic 6 relates to financial markets; Topic 5 relates to economic activity; and Topics 8, 11, and 12 relate to US national issues.

stock market has been discussed in Sect. 2 and the studies presented in [6,16] show that economic activity is a significant factor in explaining equity price movements. The impact of national issues within latent Topics 8, 11 and 12 on asset prices has been examined across a number of studies, including those with a focus on: terrorism and war [9,13,37]; man-made disasters [7,25]; environmental issues [22,26]; and innovation [18,27].

3.3   BERT for Sentiment Classification

BERT is a language model with an architecture defined by stacking the encoder units from a Transformer[3]. The significant technical development of the BERT algorithm is the use of bidirectional training in a language modelling application. Previous NLP models considered text sequences either from left-to-right or right-to-left or as a combination. The bidirectional training of BERT, where the features of the input sequence are assessed simultaneously, allows for more contextually meaningful word embeddings. That is, BERT is able to explicitly integrate both global and local context into the vector representation of a text.

A second key feature of the BERT algorithm is the use of Masked Language Modelling (MLM). MLM is a cloze test, or a fill-in-the-blank task, where the context words surrounding a masked token are used to predict the token. The prediction is made with a softmax layer over the vocabulary following the last encoder layer. The MLM methodology presents an efficient method to train a language model in a self-supervised setting.

The BERT model published by Devlin et al. [11] has two pre-trained versions: $BERT_{BASE}$, with 12 encoder layers, a hidden size of 768, 12 multi-head attention heads and 110M parameters in total; and $BERT_{LARGE}$, with 24 encoder layers, a hidden size of 1024, 16 multi-head attention heads and 340M parameters. Both of these models are pre-trained on the Books Corpus [47] with 800M words and English Wikipedia, which is comprised of 2,500M words.

Sentiment Classification. The first token of every BERT input sequence is a classification token [CLS] and the final hidden state corresponding to this token forms an aggregate representation of the sequence. That is, the final hidden state of the [CLS] token encodes the meaning of the complete input sequence. Sentiment classification is then conducted by adding a classifier layer after the final hidden state of the [CLS] token. This classifier layer is comprised of a feed-forward neural network and a softmax function. A labelled set of sentences or phrases is used to train the classifier network and fine-tune the BERT model for sentiment classification tasks.

In this study BERT has been fine-tuned for the sentiment classification of financial markets, economic activity, and US national news articles. The fine-tuning of BERT to each of these news topics, in order to optimise sentiment

---

[3] The Transformer [45] is a deep learning model that, like recurrent neural networks, is designed to handle sequential data. However, the Transformer does not feature the same network structure as an RNN and therefore does not require that the sequential data be processed in order.

quantification, was achieved using domain specific annotated data sets. The sentiment across the selected news topics in the WSJ was then estimated with the fine-tuned BERT models, considering the headline and summary sentence or sentences of each article. The following paragraphs elaborate on the WSJ articles analysed in this study and the fine-tuning of BERT to each news topic.

Sentiment Classification of Financial Market News. News articles identified as financial markets news were taken as those published in the 'Markets' section of the WSJ. Over the analysis period from January 1, 2015 to January 1, 2021 there were 63,737 articles published in this section.

For the sentiment classification of the WSJ financial news articles the domain specific language model presented in Araci [3] was used. The language model presented in [3] is the pre-trained $BERT_{BASE}$ model further trained on a financial news subset of the Reuters TRC2 corpus. This additional pre-training of the $BERT_{BASE}$ model in the target domain is performed to improve the model's classification performance [23].

To train the classifier network and fine-tune the financial news BERT model data from the Financial Phrase Bank [32] was used. The Financial Phrase Bank provides a collection of approximately 5000 phrases/sentences sampled from financial news texts and company press releases. These texts have been tagged as positive, negative, or neutral by a group of 16 annotators with business education backgrounds. A 2,265 element subset of the complete phrase bank was employed in this study, where the sentiment classification of the phrases is unanimous across the annotators.

Sentiment Classification of Economic News. In this study we focused on news articles related to economic activity, identified as Topic 5 in Table 1. News articles from the 'Economy' section of the WSJ were therefore selected which primarily contain information relating to jobs figures, unemployment figures, and construction and manufacturing activity. News items related principally to monetary and fiscal policy, identified as Topic 7 in Table 1, were disregarded, as were news items not directly related to the US economy. Between January 1, 2015 and January 1, 2021 the WSJ published 4,186 articles on the topic of US economic activity.

As with the financial news articles, the domain specific language model of Araci [3] was used in the sentiment analysis of the economic news items. There is, however, no high-quality, publicly available labelled data set for economic news articles. We therefore manually labelled 800 economic news headlines and article summaries from the WSJ as having either positive, negative or neutral sentiment[4]. These labelled articles were then used to train the classifier network and fine-tune the economic news BERT model.

---

[4] The economic and national news items were labelled with regard to the tone or emotion–namely positive, negative or neutral–expressed by the article. The subject matter of the article did not effect the sentiment label that was given to it.

Sentiment Classification of US National News. The final news topic considered in this study is broadly labelled in the WSJ as 'US News'. This news section encompasses Topics 8, 11 and 12 in Table 1. In the interest of clearly defining US national news items we considered articles that focused on: criminal justice; national defence; education; health care; racial equality; innovation; and infrastructure. Global issues that impact US interests were also considered, namely: nuclear warfare; biological warfare; pandemics; and climate change. Over the period from January 1, 2015 to January 1, 2021 there were 2,449 articles published in the WSJ meeting the definition of US national news.

The language model used for the sentiment analysis of the US national news articles was the BERT$_{BASE}$ model, as described in [11]. To train the classifier network and fine-tune the US national news BERT model we used 800 labelled news headlines and article summaries. These news headlines and article summaries were selected from the US News section of the WSJ and were manually labelled as having positive, negative or neutral sentiment (See footnote 4).

Implementation. The BERT models were fine-tuned with a learning rate of $2e^{-5}$, a mini-batch size of 32, a warm-up proportion of 0.1 and a maximum sequence length of 128 tokens. To account for label imbalance in the training data the training sets were stratified, that is, data was sampled from the training sets to achieve approximately equal class frequencies. Each model was trained on a Google Colab instance with a single NVIDIA Tesla T4 GPU with 16 GiB of memory and the models were evaluated with a 10-fold cross validation procedure.

3.4 Times Series Analysis of Sentiment Scores

Analysis of Daily Sentiment Scores. The econometric approach used to examine the impact of daily sentiment on asset prices is similar to that presented in Garcia [17]. To formalise the relationship between stock returns and news sentiment the following model has been employed

$$r_t^d = \alpha + \beta_1 N_t^f + \beta_2 N_t^e + \beta_3 N_t^n + \gamma r_{t-1}^d + \epsilon_t. \tag{3}$$

The dependent variable r_t^d is the daily open-to-close log return on the S&P 500 index. The variable N_t^i is the news sentiment score from topic i published in the six-hour period before the market opens. The three topics, as detailed in Sect. 3.2, are financial markets f, economic activity e, and US national news n. The constant term in the regression is denoted with α and ϵ_t is a residual term.

The lagged return r_{t-1}^d in Eq. 3 controls for the autocorrelation that may appear between the sentiment in an article published on day t and the previous day's return on the market. That is, the lagged return accounts for news sentiment on day t related to market activity from the previous day. If no sentiment score was recorded on a given day, indicating that an article was not published before the market opened, a neutral sentiment value was recorded. The significance of the regressors in Eq. 3 have been assessed with Newey-West [34] heteroskedasticity and autocorrelation consistent standard errors.

Analysis of Monthly Sentiment Scores. In addition to the analysis of daily sentiment scores the influence of longer-term sentiment on asset prices has been assessed. The relationship between monthly stock returns and an aggregate measure of monthly sentiment has been formalised with the univariate regression

$$
\begin{aligned}
r_t^m =&\alpha + \beta_1 N_{t-j\to t-1}^f + \beta_2 N_{t-j\to t-1}^e + \beta_3 N_{t-j\to t-1}^n \\
&+ \gamma r_{t-j\to t-1}^m + \epsilon_t \qquad \text{for } j = 1,2,3
\end{aligned}
\tag{4}
$$

where r_t^m is the monthly return on the market and ϵ_t is a residual term. The variable $N_{t-j\to t-1}^i$ is an average of the news sentiment within topic i over the preceding j months. By utilising monthly returns instead of a moving average there is no significant overlap in the sample (the importance of this is discussed in [5]). The significance of the regressors in Eq. 4 have been assessed with Newey-West [34] standard errors.

4   Results

The evaluation metrics of the fine-tuned BERT models are presented in Table 2. The models have been evaluated across the 10 validation sets held-out during the k-folds cross-validation procedure. The low cross-entropy loss values together with the high accuracy scores show that the models fit the training data well. The financial news BERT model achieves results similar to those published in Aracia [3], which is not surprising as a similar methodology has been employed and the same data sets have been utilised. The financial news BERT is also the best performing model. This is likely due the quality of the data set used to fine-tune the model, which is superior to the smaller, less refined data sets used to fine-tune the economic and national news BERT models.

Table 2. Evaluation metrics of the BERT models.

Model	Loss[a]	Accuracy[b]
BERT (Financial markets)	0.0285	0.9492
BERT (Economic activity)	0.0670	0.7999
BERT (US national news)	0.1922	0.7135

[a]The mean cross entropy loss measured on the validation sets.
[b]The ratio of correct predictions to the total number of predictions measured on the validation sets.

The fine-tuned BERT models presented in Table 2 were subsequently used to estimate the sentiment across the three selected news topics in the WSJ corpus. A 60-day moving average of these sentiment scores and their relationship to the S&P 500 index is graphically presented in Fig. 2. From the figure there appears to be a strong correlation between news sentiment and S&P 500, especially within the financial markets and economic news topics.

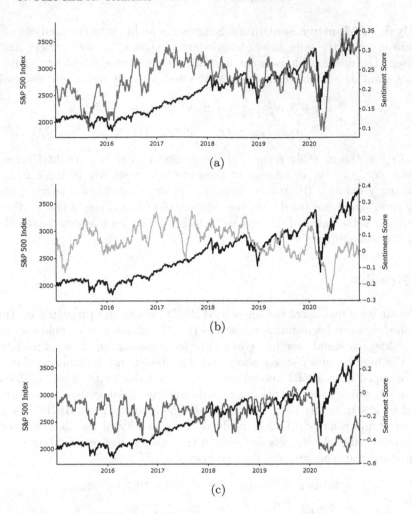

Fig. 2. Relationship between the S&P 500 and the 60-day moving average of sentiment in (a) financial markets, (b) economic, and (c) US national news articles published by the WSJ. The classification of the news articles is defined in Sect. 3.3.

The most significant feature in Fig. 2 is the sharp drop in all three sentiment indices together with a decline in the S&P 500 index in early 2020. This coincides with the escalation of the COVID-19 crises. The financial news sentiment and the S&P 500 then recover in mid-2020 and move past the pre-pandemic highs in late 2020. The economic sentiment index recovers partially in mid-2020, which may indicate the staggered reopening of the US economy. The national news sentiment is not seen to recover significantly in 2020.

Table 3 presents the parameter values from the model specification in Eq. 3, together with the Newey-West [34] t-statistics. From the results it can be seen that the sentiment of financial news published in the six hours before the market opens is predictive of same-day returns on the S&P 500 index. This effect is statistically significant at the 1% confidence level and is robust to lagged returns. The result implies that there is positive feedback from financial news to the stock market. In contrast to financial news, the sentiment in economic and US national news stories published before the market opens is not seen to be predictive of daily stock returns.

Table 3. Daily feedback from news content to the S&P 500.

	α, β, γ	t-stat
Constant α	−0.0007	−2.683
Financial news β_1	0.0030	7.790
Economic news β_2	−0.0049	−1.391
US News β_3	−0.0049	−1.793
Lagged Return γ	−0.1361	−2.077

This table reports the estimated coefficients from the model
$$r_t^d = \alpha + \beta_1 N_t^f + \beta_2 N_t^e + \beta_3 N_t^n + \gamma r_{t-1}^d + \epsilon_t.$$
Each of the variables are defined in Sect. 3.4. The sample comprises 1,508 trading days. The t-stats are computed using Newey-West [34] standard errors with 6 lags.

The second relationship examined in this paper assesses the feedback from longer-term sentiment to financial market returns. This relationship is estimated with the specification given in Eq. 4, regressing the sentiment scores in the preceding month(s) on the return in month t. The parameters of the estimated model are presented in Table 4.

In contrast to daily sentiment, the results in Table 4 illustrate that financial news sentiment is not predictive of longer-term returns on the S&P 500 index. However, there appears to be a negative relationship between long-term economic sentiment and monthly stock returns. This relationship is significant at the 1% confidence level when economic sentiment is considered over the preceding two- and three-month periods. We speculate that this negative relationship may be due to economic downturns leading to fiscal stimulus and lower interest rates, which in turn push up stock prices. A similar relationship between daily returns and macro-economic news announcements has been presented in [6, 16].

Table 4. Monthly feedback from news content to the S&P 500.

	$j = 1$		$j = 2$		$j = 3$	
	α, β, γ	t-stat	α, β, γ	t-stat	α, β, γ	t-stat
Constant α	0.0007	0.029	0.0341	0.849	0.0075	0.254
Financial news β_1	0.1218	1.147	−0.0734	−0.386	0.0937	0.648
Economic news β_2	−0.1066	−1.753	−0.1393	−2.637	−0.2087	−3.535
US news β_3	0.0943	0.862	−0.0608	−0.543	−0.0466	−0.433
Lagged return γ	−0.3303	−2.247	−0.1168	−0.827	−0.2715	−3.265

This table reports the estimated coefficients from the model
$r_t^m = \alpha + \beta_1 N_{t-j\to t-1}^f + \beta_2 N_{t-j\to t-1}^e + \beta_3 N_{t-j\to t-1}^n + \gamma r_{t-j\to t-1}^m + \epsilon_t$ for $j = 1, 2, 3$.
Each of the variables are defined in Sect. 3.4. The sample comprises 71 months and the t-stats are computed using Newey-West [34] standard errors with 3 lags.

5   Conclusion

Several recent papers have utilised BERT for sentiment classification in the finance domain. These papers have not, however, applied the model to extensive real-world data sets or provided a robust assessment of whether the estimated sentiment scores affect asset prices. In this paper a fine-tuned BERT model has therefore been used to classify the sentiment in over 63,000 financial market related articles published in the WSJ between January 1, 2015 and January 1, 2021. An econometric analysis of the sentiment scores has illustrated that sentiment in financial news articles published in the six-hour period before trading begins is predictive of same-day returns on the S&P 500 index. The longer-term sentiment in financial news is, however, not predictive of monthly returns.

In addition to financial news the analysis has been extended to WSJ articles that relate to economic activity as well as US national issues. This necessitated the fine-tuning of BERT to these additional news topics. The results indicate that there is a statistically significant negative relationship between long-term economic sentiment and monthly returns on the S&P 500 index. We speculate that this negative relationship may be due to economic downturns prompting fiscal stimulus and lower interest rates, which in turn push up stock prices.

While the methodology presented in this paper has provided a more detailed understanding of the relationship between news media and asset prices there are still limitations. Firstly, the quality and quantity of the data used to fine-tune the BERT models for economic and national news items may be improved, increasing the accuracy of the models. Furthermore, the number of news topics analysed and the analysis period may be extended in future studies.

References

1. Algaba, A., Ardia, D., Bluteau, K., Borms, S., Boudt, K.: Econometrics meets sentiment: an overview of methodology and applications. J. Econ. Surv. **34**(3), 512–547 (2020)
2. Antweiler, W., Frank, M.Z.: Is all that talk just noise? the information content of internet stock message boards. J. Finance **59**(3), 1259–1294 (2004)
3. Araci, D.: Finbert: Financial sentiment analysis with pre-trained language models. arXiv preprint arXiv:1908.10063 (2019)
4. Blei, D.M., Ng, A.Y., Jordan, M.I.: Latent dirichlet allocation. J. Mach. Learn. Res. **3**, 993–1022 (2003)
5. Boudoukh, J., Israel, R., Richardson, M.: Long-horizon predictability: a cautionary tale. Finan. Anal. J. **75**(1), 17–30 (2019)
6. Boyd, J.H., Hu, J., Jagannathan, R.: The stock market's reaction to unemployment news: why bad news is usually good for stocks. J. Finance **60**(2), 649–672 (2005)
7. Capelle-Blancard, G., Laguna, M.A.: How does the stock market respond to chemical disasters? J. Environ. Econ. Manage. **59**(2), 192–205 (2010)
8. Chen, C.C., Huang, H.H., Chen, H.H.: Issues and perspectives from 10,000 annotated financial social media data. In: Proceedings of the 12th Language Resources and Evaluation Conference, pp. 6106–6110 (2020)
9. Chesney, M., Reshetar, G., Karaman, M.: The impact of terrorism on financial markets: an empirical study. J. Bank. Finance **35**(2), 253–267 (2011)
10. Das, S.R., Chen, M.Y.: Yahoo! for amazon: sentiment extraction from small talk on the web. Manage. Sci. **53**(9), 1375–1388 (2007)
11. Devlin, J., Chang, M.W., Lee, K., Toutanova, K.: Bert: Pre-training of deep bidirectional transformers for language understanding. arXiv preprint arXiv:1810.04805 (2018)
12. Dougal, C., Engelberg, J., Garcia, D., Parsons, C.A.: Journalists and the stock market. Rev. Finan. Stud. **25**(3), 639–679 (2012)
13. Eldor, R., Melnick, R.: Financial markets and terrorism, Eur. J. Polit. Econ. **20**(2), 367–386 (2004)
14. Fama, E.F.: Efficient capital markets: a review of theory and empirical work. J. Finance **25**(2), 383–417 (1970)
15. Friedman, M.: Essays in Positive Economics. University of Chicago Press, Chicago (1953)
16. Funke, N., Matsuda, A.: Macroeconomic news and stock returns in the united states and Germany. German Econ. Rev. **7**(2), 189–210 (2006)
17. Garcia, D.: Sentiment during recessions. J. Finance **68**(3), 1267–1300 (2013)
18. Garleanu, N., Panageas, S., Yu, J.: Technological growth and asset pricing. J. Finance **67**(4), 1265–1292 (2012)
19. Gentzkow, M., Kelly, B., Taddy, M.: Text as data. J. Econ. Lit. **57**(3), 535–74 (2019)
20. Gino, F., Brooks, A.W., Schweitzer, M.E.: Anxiety, advice, and the ability to discern: feeling anxious motivates individuals to seek and use advice. J. Pers. Soc. Psychol. **102**(3), 497 (2012)
21. Gurun, U.G., Butler, A.W.: Don't believe the hype: local media slant, local advertising, and firm value. J. Finance **67**(2), 561–598 (2012)
22. Hamilton, J.T.: Pollution as news: media and stock market reactions to the toxics release inventory data. J. Environ. Econ. Manage. **28**(1), 98–113 (1995)

23. Howard, J., Ruder, S.: Universal language model fine-tuning for text classification. arXiv preprint arXiv:1801.06146 (2018)
24. Jegadeesh, N., Wu, D.: Word power: a new approach for content analysis. J. Finan. Econ. **110**(3), 712–729 (2013)
25. Kaplanski, G., Levy, H.: Sentiment and stock prices: the case of aviation disasters. J. Financ. Econ. **95**(2), 174–201 (2010)
26. Klassen, R.D., McLaughlin, C.P.: The impact of environmental management on firm performance. Manage. Sci. **42**(8), 1199–1214 (1996)
27. Kung, H., Schmid, L.: Innovation, growth, and asset prices. J. Finance **70**(3), 1001–1037 (2015)
28. Larsen, V.H., Thorsrud, L.A.: The value of news for economic developments. J. Econometrics **210**(1), 203–218 (2019)
29. Loughran, T., McDonald, B.: When is a liability not a liability? textual analysis, dictionaries, and 10-ks. J. Finance **66**(1), 35–65 (2011)
30. Loughran, T., McDonald, B.: Textual analysis in accounting and finance: a survey. J. Acc. Res. **54**(4), 1187–1230 (2016)
31. Loughran, T., McDonald, B.: Textual analysis in finance. Ann. Rev. Financ. Econ. **12**, 357–375 (2020)
32. Malo, P., Sinha, A., Korhonen, P., Wallenius, J., Takala, P.: Good debt or bad debt: detecting semantic orientations in economic texts. J. Assoc. Inf. Sci. Technol. **65**(4), 782–796 (2014)
33. Nardo, M., Petracco-Giudici, M., Naltsidis, M.: Walking down wall street with a tablet: a survey of stock market predictions using the web. J. Econ. Surv. **30**(2), 356–369 (2016)
34. Newey, W.K., West, K.D.: A simple, positive semi-definite, heteroskedasticity and autocorrelation consistent covariance matrix. Econometrica J. Econometric Soc. **55**(3), 703–708 (1987)
35. Řehřek, R., Sojka, P., et al.: Gensim—statistical semantics in python. Retrieved from genism. org (2011)
36. Renault, T.: Intraday online investor sentiment and return patterns in the us stock market. J. Bank. Finance **84**, 25–40 (2017)
37. Rigobon, R., Sack, B.: The effects of war risk on us financial markets. J. Bank. Finance **29**(7), 1769–1789 (2005)
38. Shapiro, A.H., Sudhof, M., Wilson, D.J.: Measuring news sentiment. J. Econometrics (2020)
39. Shiller, R.J.: Narrative Economics. Princeton University Press, Princeton (2020)
40. Sievert, C., Shirley, K.: Ldavis: a method for visualizing and interpreting topics. In: Proceedings of the Workshop on Interactive Language Learning, Visualization, and Interfaces, pp. 63–70 (2014)
41. Solomon, D.H., Soltes, E., Sosyura, D.: Winners in the spotlight: media coverage of fund holdings as a driver of flows. J. Financ. Econ. **113**(1), 53–72 (2014)
42. Sousa, M.G., Sakiyama, K., de Souza Rodrigues, L., Moraes, P.H., Fernandes, E.R., Matsubara, E.T.: Bert for stock market sentiment analysis. In: 2019 IEEE 31st International Conference on Tools with Artificial Intelligence (ICTAI), pp. 1597–1601. IEEE (2019)
43. Tetlock, P.C.: Giving content to investor sentiment: the role of media in the stock market. J. Finance **62**(3), 1139–1168 (2007)
44. Tetlock, P.C.: Information transmission in finance. Annu. Rev. Financ. Econ. **6**(1), 365–384 (2014)
45. Vaswani, A., et al.: Attention is all you need. In: Advances in Neural Information Processing Systems, pp. 5998–6008 (2017)

46. Yang, Y., Uy, M.C.S., Huang, A.: Finbert: A pretrained language model for financial communications. arXiv preprint arXiv:2006.08097 (2020)
47. Zhu, Y., et al.: Aligning books and movies: towards story-like visual explanations by watching movies and reading books. In: Proceedings of the IEEE International Conference on Computer Vision, pp. 19–27 (2015)

Investigation of Topic Modelling Methods for Understanding the Reports of the Mining Projects in Queensland

Yasuko Okamoto[1], Thirunavukarasu Balasubramaniam[2(✉)], and Richi Nayak[2]

[1] RecordPoint, Sydney, Australia
[2] School of Computer Science and Centre for Data Science,
Queensland University of Technology, Brisbane, Australia
{thirunavukarasu.balas,r.nayak}@qut.edu.au

Abstract. In the mining industry, many reports are generated in the project management process. These past documents are a great resource of knowledge for future success. However, it would be a tedious and challenging task to retrieve the necessary information if the documents are unorganized and unstructured. Document clustering is a powerful approach to cope with the problem, and many methods have been introduced in past studies. Nonetheless, there is no silver bullet that can perform the best for any types of documents. Thus, exploratory studies are required to apply the clustering methods for new datasets. In this study, we will investigate multiple topic modelling (TM) methods. The objectives are finding the appropriate approach for the mining project reports using the dataset of the Geological Survey of Queensland, Department of Resources, Queensland Government, and understanding the contents to get the idea of how to organise them. Three TM methods, Latent Dirichlet Allocation (LDA), Nonnegative Matrix Factorization (NMF), and Nonnegative Tensor Factorization (NTF) are compared statistically and qualitatively. After the evaluation, we conclude that the LDA performs the best for the dataset; however, the possibility remains that the other methods could be adopted with some improvements.

Keywords: Document clustering · Topic model · Document management · Information retrieval · Mining industry

1 Introduction

For the mining industry, documentation is one of the key components. The state governments in Australia publish the framework for the mining project management and instruct the documentation in detail. Thus, a large number of documents are generated as the project milestone products every year. These documents comprise feasibility studies, risk location, change management, etc., and they contain knowledge for future success. Therefore, building the document management and information retrieval system is a crucial task in the industry.

© Springer Nature Singapore Pte Ltd. 2021
Y. Xu et al. (Eds.): AusDM 2021, CCIS 1504, pp. 196–208, 2021.
https://doi.org/10.1007/978-981-16-8531-6_14

Regarding the mining projects in Queensland, the Geological Survey of Queensland (GSQ), Department of Resources, Queensland Government has collected the project reports form mining companies over the past hundred years. The collection is quite complex since it consists of more than 100,000 reports submitted by over 5,000 companies. Furthermore, the documents are currently unstructured and unorganised, which means the documents are not searchable, and information retrieval from such a collection is a tedious task.

Document clustering is a text mining approach to cope with the problem by grouping document based on their topic and similarities by automatically finding patterns and characteristics from data itself without any predefined data labels. With its capability, document clustering has been widely applied to understand the collection of documents and build a robust search engine [1]. The paper will focus on one of the document clustering techniques, the topic modelling (TM). TM discovers topics hidden in the collection as well as describes the association between the topics and each document. The advantage is that it does not only group the documents but tells the contents of each group simultaneously. It has been studied by many researchers, and the past research has developed multiple algorithms, such as probabilistic latent semantic analysis (PLSA) [2], latent Dirichlet allocation (LDA) [3], and non-negative matrix factorization (NMF) [4]. In recent years, the studies on the application of TM methods have been also active, especially for the domain where the expertise is required, for example, bio-informatic, healthcare, and medical [5–8].

While a variety of methods and use cases have been introduced, the nature of text mining is domain-specific. It means that the appropriate approach can vary depending on the data types and purposes. Therefore, the exploratory studies are continuously needed for the application in the real-world. Thus, this study investigates the performance of several TM methods and explores the best approach for GSQ's documents. Besides, we expect that the grouping results obtained from the experiment would be a help to understand the contents of the collection and analyse the trends of project and business. Furthermore, these findings could derive the category labels and the summary of the contents of each group, which can be essential factors and materials for future studies on building a robust document management system for GSQ.

The rest of this paper is structured as follows. Section 2 elaborates the recent studies on TM methods and applications and identifies the problem space to be addressed in the study. Section 3 describes the proposed methods. Algorithms, datasets, and evaluation methods applied in the experiment will be explained. The results of TM will be observed and analysed in Sect. 4. Finally, Sect. 5 provides the conclusion and suggestions for future studies.

2   Related Works

2.1   Document Representation Model

The basic concept of TM methods is taking the numeric (document x term) vectors that represent the documents as input and converting them into (topic x term) vectors and (document x topic) vectors. How documents should be represented in the vector space, or the document representation model is one of the important parameters for TM methods. This sub-section briefly reviews and introduces several models. The most classic one is Bag-of-words (BoW) model where documents are represented as the collection of words, and the word order information in the document is ignored. In the simple word count vector, for example, a document is a sequence of N words denoted by $D = \{tf_1, tf_2, ..., tf_N\}$, where tf_i is the frequency of term i in document D. The term-frequency-inversed-document-frequency (TFIDF) is a popular scheme of the BoW model. It adds the importance of the term in the collection to the count vector. However, there was a common awareness of the limitation of these approaches among the researchers: the document representation models lose the semantics and relationships among words. The powerful technique to address this problem is word-embedding. In 2013, the research team at Google advocated a new approach, called Word2Vec [9], which involved two-layers neural networks and embedded each word to a numeric vector space. The technique allows to calculate the semantic similarity between words, and a document can be represented by a vector incorporating the relationship between the words using the average values of Word2Vec [10]. The technique was further developed and expanded to the document scale, which is called Doc2Vec [11].

2.2   Recent Studies of Topic Modelling

Many TM methods have been developed in past studies, and they can be categorised into two types: the probabilistic model and the non-probabilistic model. The examples of former are PLSA [2] and LDA [3]. Both calculate the probability of the word appearing in the different topics, P(word—topic), and the probability of the topics in the document P(topic—document). Latent Semantic Analysis (LSA) [12] and NMF [4] are categorised in non-probabilistic approaches that are algebraic approaches using matrix factorization. While there are several methods as above mentioned, LDA and NMF are the most popular techniques for industrial application [13–15].

At the same time, researchers keep investigating better TM approaches using different datasets. Studies of Anantharaman, et al., Ray, et al., and Chehal, et al. compared the existing methods explained above, and the conclusions of the studies show that the best methods are different depending on the datasets [16–18]. Furthermore, the researchers sometimes modified the exiting methods to adjust them according to the datasets or purposes. Wang, et al. introduced the Multi-Attribute LDA that can consider additional document attributes other than terms when mining the topics [19]. With the extended LDA, they concluded that hot topics in the microblog could be found more properly using not only texts, but the post time and hashtag information compared to the normal LDA. Shi, et al. proposed a semantics-assisted non-negative matrix factorization (SeaNMF) [20]. While the original NMF uses the BoW document model, they argued that incorporating the semantic correlations between words discovered more coherent topics compared with other methods such as normal LDA and NMF.

The recent studies imply that the best approach varies depending on the datasets, and the comparative studies are required when applying TM to the new datasets while LDA and NMF are the most popular methods. Moreover, the methods sometimes should be modified according to the purpose. Therefore, this research investigates multiple methods to answer What TM approach is appropriate to analyse the contents of GSQ's mining project reports?

3   Topic Modelling Approaches

Given a large number of documents and companies who submitted the reports, the study focuses on two specific objectives: grouping documents by contents similarities, and grouping companies based on their reports. For the former objective, LDA and NMF are applied considering their popularities and characteristics that worked better on the long texts as seen in the recent studies. To achieve the latter objective, 3-dimensional TM that can take the company information of documents into account.

3.1   Two-Dimensional Topic Modelling: LDA and NMF

Both algorithms accept the input of (document x term) matrix and generate two matrices that are (document x topic) matrix and (topic x term) matrix. The former denotes how each document is associated with each topic, and the latter can be considered as topics list discovered from the collection. In the study, each document is assigned to one topic group using the (document x topic) matrix. More precisely, document i is assigned to topic x if $x = \text{argmax}_j v_{ij}$, where v_{ij} each element of (document x topic) matrix representing document i is associated with topic j. This assigning method is referred to as the argmax method in the rest of the paper.

LDA is a probabilistic model introduced by Blei et al. [3], and the basic concept is assuming that each document consists of multiple topics, where each topic is a probability distribution over words. Assume K is the number of topics, N is the number of words in the document, M is the number of documents in the collection, α is the parameter representing the Dirichlet prior for the per-document topic distribution, β is the parameter representing the Dirichlet for the per-topic word distribution, $\varphi(k)$ is the word distribution for topic k, $\theta(i)$ is the topic distribution for document i, $z(i,j)$ is the topic assignment for $w(i,j)$, $w(i,j)$ is the word j in document i. The aim is to learn φ (topic x term) matrix, and θ (document x topic) matrix. α, β, and K should be specified by the user. In this study, exploring the best K is focused on. Therefore, α and β are set to the default values of LDA function of scikit-learn package, which are $1/K$. The range of K is explained in the coming section of the paper.

NMF is a matrix factorization method. Given the original matrix $\mathbf{X}$ (document x term), two matrices can be obtained $\mathbf{W}$ (topic x term) and $\mathbf{H}$ (topic x document), such that $\mathbf{X} = \mathbf{WH}$. By taking advantage of the fact that $\mathbf{X}$

is non-negative, the two matrices $\mathbf{W}$ and $\mathbf{H}$ are optimised over the following objective function using Frobenius norm:

$$\frac{1}{2}\|\mathbf{X} - \mathbf{WH}\|^2 = \sum_{i=1}^{n} \sum_{j=1}^{m} (\mathbf{X}_{ij} - (\mathbf{WH}_{ij}))^2 \tag{1}$$

In the function, the error of reconstruction between $\mathbf{X}$ and the product of $\mathbf{W}$ and $\mathbf{H}$ is measured. Thus, $\mathbf{W}$ and $\mathbf{H}$ are updated iteratively using the following update function until convergence.

$$\mathbf{W} \leftarrow \mathbf{W} \frac{\mathbf{XH}}{\mathbf{WHH}} \tag{2}$$

$$\mathbf{H} \leftarrow \mathbf{H} \frac{\mathbf{WX}}{\mathbf{WWH}} \tag{3}$$

In the study, only K is tuned using multiple values. The other parameters other than the method to initialise the procedure were set to the default values of NMF function of scikit-learn package. For the initialisation method, Nonnegative Double Singular Value Decomposition (NNDSVD) is used since it is considered better for sparse data [21].

3.2   Three-Dimensional (3D) Topic Modelling: Tensor Clustering

3D TM can accept a 3D matrix while LDA and NMF accept a 2D matrix. Hence, it can group data with two criteria simultaneously. In the study, a 3D matrix (document x company x term) matrix is fed into the algorism, and it generates three 2D matrices that are (document x topic) matrix, (term x topic) matrix and (company x topic) matrix. The third matrix describes how each company is associated with each topic. Thus, the method enables to group documents and group companies at once. The method is referred to as the tensor clustering in the paper. The applied method in this study is the Saturating Coordinate Descent (SaCD) that is an improved method of Nonnegative Tensor Factorization (NTF) [22]. The NTF model is formulated as follows:

$$\min_{\mathbf{U},\mathbf{V},\mathbf{W} \geq 0} f(\mathbf{U}, \mathbf{V}, \mathbf{W}) = \|\boldsymbol{\mathcal{X}} - \mathbf{U}, \mathbf{V}, \mathbf{W}\|^2 \tag{4}$$

where $\boldsymbol{\mathcal{X}}$ is a 3D tensor, which is (document $\times$ term $\times$ company) tensor here, and $\mathbf{U}, \mathbf{V}, \mathbf{W}$ are the factor matrices for $\boldsymbol{\mathcal{X}}$. SaCD makes this algorithm faster and more scalable applying the updated method that reduces the calculation in the learning process. After obtaining (document $\times$ topic) matrix and (company $\times$ topic) matrix, each document and each company is assigned to one topic using the argmax method in the same manner as 2D TM.

3.3 Dataset

This section discusses the datasets used in the experiment. We use only the reports that had been converted into Microsoft (MS) Word format and stored. Besides, the report can consist of more than one documents. In that case, only the body of the report is used.

Given that the collection is quite large, and categorised by mining industries, such as coal, mineral, petroleum, it is assumed that performing TM methods onto the subsets of each category would provide new findings rather than performing TM onto the whole collection. Therefore, this study focus on performing TM methods on the target subsets. The target subsets are chosen by the following steps: According to the metadata file, more than 60% of the reports were submitted in the last two decades. Therefore, the target period is set to 2000–2020. Next, by aggregating the reports of this period, the three dominants categorise are found: Coalbed Methane, Gold and Coal. In this paper we only report the outputs of Coal.

Moreover, the reports of each category have the information about report types, such as well completion report, annual report, and final report. Hence, the target was further narrowed down by report type. In the coal, the annual report is the main report type that occupies 80%.

After setting the target period, categories, and report types, the number of reports available in MS word format is counted. The data is also cleaned by the company name identification and removing the documents that had not been converted into MS word format properly.

3.4 Data Preprocessing

Text Extraction. The process was to parse the reports in Microsoft word format and extracting the body text with python programming. Tab and newline were removed here. At the end of the process, the data frame that contains report - body text (a string) information is obtained.

Text Preprocessing. The text pre-processing is the process to clean the documents using the natural language processing (NLP) methods before converting them into the numeric data. The details are as follows:

1. Tokenisation: In the tokenisation process, the texts in documents are split text into words and lowercased using spaCy, the NLP package in Python. Punctuations, special characters, words with less than three characters are removed after tokenisation.
2. Stop-word removal: Stop-words are the terms that have little meaning and occur in the document with high frequencies, such as delimiters and prepositions. The terms are removed before performing TM. The stop-words list from the NLTK corpus is applied. Besides, the additional 13 stop-words are chosen, which are: appendix, area, Australia, fax, figure, ltd, map, page, phone, project, report, year, within.

3. Stemming: Stemming is the process to uniform the words in the different morphological forms. For instance, the term "program" can be different forms such as "programs", "programmer", "programmers", "programming" will be uniformed into "program. The Porter Stemming Algorithm advocated by Martin Porter in 1980 is applied using the NLTK package.

Data Transformation. The documents transformation process is to convert the documents list into a numeric matrix to feed the TM algorithm. As discussed in Sect. 2, there are several document representation models; however, following the past studies that introduced LDA and NMF for TM [3,4], the BoW model is adopted. More specifically, each document is vectorized with the term-frequency (TF) for LDA and TFIDF values for NMF. As explained briefly in Sect. 2, TFIDF is the weighting schema that adds the importance of the term in the collection.

All documents in the dataset are converted into vectors in the same manner. Thus, the 312 documents of Coal dataset with 8,549 unique terms, are converted into 2D TF matrix and TFIDF matrix with the shape of (312, 8,549) for LDA and NMF. For the 3D tensor clustering, TF values are adopted, and the company information is added to the (document × term) TF matrix. Therefore, the Coal dataset with 76 companies is converted into the 3D matrix with the shape of (312, 76, 8549) for the tensor clustering.

3.5   Evaluation Measures

Silhouette analysis is one of the major evaluation methods of clustering approach and can be used to analyse the separation distance between the clusters. In the method, the silhouette coefficient of each sample is calculated, which indicate how much the sample is far away from the neighbouring clusters with the following equation.

$$S_i = (b_i - a_i)/max(a_i, b_i) \tag{5}$$

where S_i the silhouette coefficient of sample i, a_i the average distance between i and all the other data points in the cluster to which i belongs, and b_i is the minimum average distance from i to all clusters to which i does not belong. The value of the silhouette coefficient is between $[-1, 1]$. The value closer to 1 denotes that the sample is far from the neighbouring clusters, and the negative value implies the sample might have been assigned to the wrong cluster. In the study, silhouette scores are calculated using (topic x document) matrix for LDA/NMF. For the tensor clustering, (topic x document) matrix is used to evaluate the grouping quality for document groups, and (topic x company) matrix is used for the quality of company groups. For each K, the average silhouette score is calculated, as well as the silhouette coefficients of all samples are plotted in the figure to observe whether the samples of each group were assigned properly. The candidate numbers for K are chosen for each dataset based on the average silhouette scores. Then, the plotting results of silhouette scores and the results of the second evaluation method, topic keywords matching, are taken into account to decide the best number for K among the candidates and the best TM methods for grouping documents.

3.6   Topic Keyword Matching

After selecting candidate Ks with the above method, for each K, the top-30 keywords of each group are observed. The top-30 keywords are obtained from (topic × term) matrix generated by TM. The 30 words are compared with the 30 terms appearing in the documents of the group most frequently. The purpose of the method is to check whether the keywords found by TM actually appear in the documents of the group.

4   Results and Discussion

The purpose of 2D TM is grouping documents. The best K and the best method were identified among the results of LDA, NMF, and the grouping documents result of the tensor clustering.

For the Coal dataset, K = 3 and 4 resulted in the highest average silhouette scores for all algorithms (Fig. 1). Furthermore, the silhouette score comparison and the keyword matching ratio comparison (Fig. 2) shows that LDA performed the best. Finally, from the facts that the keyword matching ratio is higher, and the silhouette analysis resulted better when K = 4, 4 was chosen as the best number of groups for Coal dataset.

With 3D TM, we aimed to group documents and companies simultaneously. Therefore, the tensor clustering method was evaluated separately from 2D TM results. Overall, it can be mentioned that there are some doubts about the results of grouping companies for all dataset because the detailed silhouette analysis indicates most of the companies are assigned into the same group, and the companies that were assigned into the different groups resulted in the negative silhouette score.

For Coal dataset, 3 and 4 could be candidates for K according to the silhouette score (Figure Fig. 1). Observing the keyword match ratio, the better result is provided when $K = 4$, which was 73%. Hence, 4 was chosen as the best K. However, it should be noted that the detailed silhouette analysis for both grouping document and grouping companies indicate that the qualities of some clusters are low, which means that the tensor clustering could not perform well for this dataset.

From the evaluation results, the best K and method were chosen for each dataset as below. The following sub-sections discuss the details of grouping results of chosen K and method.

4.1   Document Group by LDA

The dataset containing 312 documents submitted by 76 companies were divided by LDA into four groups: 25 documents of 6 companies in topic #0, 253 documents of 69 companies in topic #1, 21 documents of 3 companies in topic #2, 13 documents of 8 companies of topic #3. What each topic represents is as follows: Topic #0 is about "volcan", "creek", "seismic", "cockenzi", "permian", topic #1 is about "explor", "seam", "basin", "measur", "mine", topic

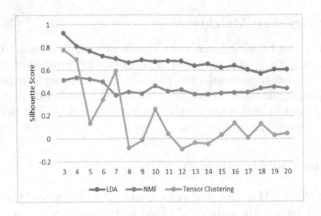

Fig. 1. Silhouette score comparison

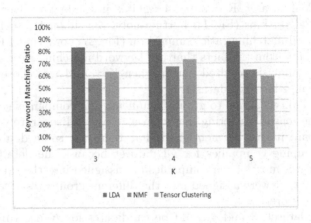

Fig. 2. Keyword matching ratio comparison with candidate number of K

#2 is about "basin", "moreton", "walloon", "surat", "basal", topic #3 is about "grey", "fresh", "grain", "medium", "siltston". Most of or all of the document of the same company were assigned to a single topic. Therefore, it can be considered the contents of reports submitted by the same company are consistent in the dataset, or each company can have a strong relationship with a single topic.

4.2   Document Group by Tensor

Four topics found by the tensor clustering are: topic #0 about "sandston", "siltston", "medium", "grey", "grain", topic #1 about "epc", "rock", "medium", "sandston", "seam", topic #2 about "thick", "grain", "grey", "sandston", "dark", topic #3 about "rock", "grain", "grey". "sandston", "clay". As seen from these keywords, the four topics are similar. However, LDA results showed that there should be several topics in the dataset. Therefore, by removing some common words such as "coal", "drill", and "geolog" as stop-words, the method could find more meaningful topics. Company groups and document groups are consistent. In both grouping results, topic #0 is occupied by Tenement Administration Services, topic #1 is occupied by Lance Grimstone & Associates Pty Ltd, topic #2 is occupied by BHP Billion Mitsubishi Alliance (BMA). The rest companies with 271 documents were assigned to topic #1. Thus, from the tensor clustering, the three companies that have unique contents in their reports were found.

4.3   Indecisiveness of Tensor Clustering

As discussed above, the statistical evaluation score of tensor clustering was low, and inconsistent results were observed. The reason for the low clustering quality was investigated under the assumption that the tensor clustering is more indecisive and the argmax method would not be appropriate. TM is originally a soft clustering technique, which does assess the association of each document with each topic. Thus, a document can belong to one or more groups. The argmax method, however, assigns each document to one topic that has the strongest association with the document. In the process, the relationships that the document might have with other topics are ignored. Hence, this method is not appropriate if the TM method is softer or more indecisive since the impact of ignored relationships would be large. For the investigation, the standard deviations of (document x topic) matrix and (company x topic) matrix are calculated and compared among TM methods. In the same manner, the average standard deviations are obtained from (document x topic) matrices of all TM methods. If the standard deviation is large, the method is evaluated to be decisive. On the other hand, the method is indecisive if the value is small.

Comparing the standard deviations from the matrices of the best K (Fig. 3), the standard deviation of the tensor clustering is smaller than LDA and NMF in all dataset. On the other hand, the value of LDA is much larger compared to other methods. It corresponds to that LDA resulted in the highest silhouette analysis score as seen in the evaluation section.

Furthermore, by comparing the standard deviations of all Ks, it is found that LDA is the most decisive, and the tensor clustering is the least decisive. Thus, it can be concluded that the argmax method is not appropriate for the tensor clustering, and the other methods need to be explored to use the outputs of tensor clustering more effectively.

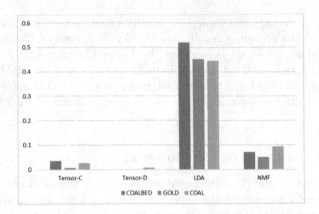

Fig. 3. Comparison of standard deviation

5   Conclusion

Comparing multiple TM methods for GSQ's reports, we conclude that LDA performs the best. However, at the same time, the need for improvement remains since it found many overlapping words over topics. For improvement, we can consider the modification of text pre-processing. Especially, some new stop-words should be added in the future experiment.

Although LDA performed the best in term of the quality of clusters, testing the other methods is still meaningful. This study involved a small dataset, but we have to expand our methods to the whole GSQ's collection in the future. Hence, faster and scalable algorithms would be more ideal. The tensor clustering method could respond to this need as it can perform much faster than LDA. Furthermore, it enables us to analyse the document from the various aspects simultaneously. However, to adopt this algorithm, it has to be improved. For the future study, we suggest the following things as the idea for the improvement: First, exploring the new assigning method other than the argmax is required. Implementing other clustering methods, such as K-Means and DBSCAN, on the output of the tensor clustering can be options. Second, the new stop-words should be added in the text pre-processing. Third, normalising the number of documents per companies can be considered.

Finally, some characteristics from dataset were found by observing the results of the best K. However, in the document clustering approach, the statistical evaluation score does not necessarily mean that the grouping results are comprehensive for human. Hence, the human interpretation of clustering results is still required. In the study, we recorded and visualised the results of all Ks using a BI tool, Tableau. We would suggest reviewing the results of other Ks in the future if the results of the best K are considered unreasonable by the experts in the industry.

Acknowledgement. This research was supported by the Geoscience Information, Geological Survey of Queensland (GSQ), Department of Resources, Queensland Government.

References

1. Gupta, P., Narang, B.: Role of text mining in business intelligence. Gian Jyoti E-J. **1**(2) (2012)
2. Hofmann, T.: Probabilistic latent semantic indexing. In: Proceedings of the 22nd Annual International ACM SIGIR Conference on Research and Development in Information Retrieval, pp. 50–57 (1999)
3. Blei, D.M., Ng, A.Y., Jordan, M.I.: Latent dirichlet allocation. J. Mach. Learn. Res. **3**, 993–1022 (2003)
4. Xu, W., Liu, X., Gong, Y.: Document clustering based on non-negative matrix factorization. In: Proceedings of the 26th Annual International ACM SIGIR Conference on Research and Development in Information Retrieval, pp. 267–273 (2003)
5. Huang, X., Zheng, X., Yuan, W., Wang, F., Zhu, S.: Enhanced clustering of biomedical documents using ensemble non-negative matrix factorization. Inf. Sci. **181**(11), 2293–2302 (2011)
6. Dantu, R., Dissanayake, I., Nerur, S.: Exploratory analysis of internet of things (IoT) in healthcare: a topic modelling & co-citation approaches. Inf. Syst. Manage. **38**(1), 62–78 (2021)
7. Feng, J., Mu, X., Wang, W., Xu, Y.: A topic analysis method based on a three-dimensional strategic diagram. J. Inf. Sci. **47**, 0165551520930907 (2020)
8. Balasubramaniam, T., Nayak, R., Luong, K., Bashar, M.A.: Identifying covid-19 misinformation tweets and learning their spatio-temporal topic dynamics using nonnegative coupled matrix tensor factorization. Soc. Netw. Anal. Min. **11**(1), 1–19 (2021)
9. Mikolov, T., Chen, K., Corrado, G., Dean, J.: Efficient estimation of word representations in vector space. arXiv preprint arXiv:1301.3781 (2013)
10. Chen, M.: Efficient vector representation for documents through corruption. arXiv preprint arXiv:1707.02377 (2017)
11. Le, Q., Mikolov, T.: Distributed representations of sentences and documents. In: International Conference on Machine Learning, pp. 1188–1196. PMLR (2014)
12. Deerwester, S., Dumais, S.T., Furnas, G.W., Landauer, T.K., Harshman, R.: Indexing by latent semantic analysis. J. Am. Soc. Inf. Sci. **41**(6), 391–407 (1990)
13. Westerlund, M., Leminen, S., Rajahonka, M.: A topic modelling analysis of living labs research. Technol. Innov. Manage. Rev. **8**(7), 40–51 (2018)
14. Zhang, T., Sahinidis, N.V., Rosé, C.P., Amaran, S., Shuang, B.: Forty years of computers and chemical engineering: analysis of the field via text mining techniques. Comput. Chem. Eng. **129**, 106511 (2019)
15. Moro, S., Pires, G., Rita, P., Cortez, P.: A text mining and topic modelling perspective of ethnic marketing research. J. Bus. Res. **103**, 275–285 (2019)
16. Anantharaman, A., Jadiya, A., Siri, C.T.S., Adikar, B.N., Mohan, B.: Performance evaluation of topic modeling algorithms for text classification. In: 2019 3rd International Conference on Trends in Electronics and Informatics (ICOEI), pp. 704–708. IEEE (2019)
17. Ray, S.K., Ahmad, A., Kumar, C.A.: Review and implementation of topic modeling in Hindi. Appl. Artif. Intell. **33**(11), 979–1007 (2019)

18. Chehal, D., Gupta, P., Gulati, P.: Implementation and comparison of topic modeling techniques based on user reviews in e-commerce recommendations. J. Ambient Intell. Humanized Comput. **12**(5), 5055–5070 (2020). https://doi.org/10.1007/s12652-020-01956-6
19. Wang, J., Li, L., Tan, F., Zhu, Y., Feng, W.: Detecting hotspot information using multi-attribute based topic model. PLoS ONE **10**(10), e0140539 (2015)
20. Shi, T., Kang, K., Choo, J., Reddy, C.K.: Short-text topic modeling via nonnegative matrix factorization enriched with local word-context correlations. In: Proceedings of the 2018 World Wide Web Conference, pp. 1105–1114 (2018)
21. Boutsidis, C., Gallopoulos, E.: SVD based initialization: a head start for nonnegative matrix factorization. Pattern Recogn. **41**(4), 1350–1362 (2008)
22. Balasubramaniam, T., Nayak, R., Yuen, C.: Efficient nonnegative tensor factorization via saturating coordinate descent. ACM Trans. Knowl. Disc. Data (TKDD) **14**(4), 1–28 (2020)

A Semi-automatic Data Extraction System for Heterogeneous Data Sources: a Case Study from Cotton Industry

Richi Nayak[1], Thirunavukarasu Balasubramaniam[1(✉)], Sangeetha Kutty[1], Sachindra Banduthilaka[2], and Erin Peterson[3]

[1] School of Computer Science and Centre for Data Science, Queensland University of Technology, Brisbane, Australia
{r.nayak,thirunavukarasu.balas,s.kutty}@qut.edu.au
[2] Redeye Apps Pvt Ltd, Brisbane, Australia
sachi.banduthilaka@redeye.co
[3] Erin Peterson Consulting, Brisbane, Australia
erin@peterson-consulting.com

Abstract. With the recent developments in digitisation, there are increasing number of documents available online. There are several information extraction tools that are available to extract information from digitised documents. However, identifying precise answers to a given query is often a challenging task especially if the data source where the relevant information resides is unknown. This situation becomes more complex when the data source is available in multiple formats such as PDF, table and html. In this paper, we propose a novel data extraction system to discover relevant and focused information from diverse unstructured data sources based on text mining approaches. We perform a qualitative analysis to evaluate the proposed system and its suitability and adaptability using cotton industry.

Keywords: Information extraction · Focused information retrieval · Automated discovery · NER · Chunking · Unstructured data · Web

1 Introduction

In the recent years, agricultural data is available in a wide variety of sources. Information overload digitising the documents has resulted in considerable benefits such as accessibility, cost savings and disaster recovery. Due to these advantages many enterprises digitise their documents online resulting in abundant data on the web and with the enterprises [1]. The digitization of newspaper articles, government policy documents, research publications and statistical reports has contributed to the creation of big data [2]. One of the challenges of the big data is the information overload due to which it is difficult for humans to sift through the information on a subject topic and identify the relevant value for the information of interest. For example, it is a challenge for a farmer or a policy officer

© Springer Nature Singapore Pte Ltd. 2021
Y. Xu et al. (Eds.): AusDM 2021, CCIS 1504, pp. 209–222, 2021.
https://doi.org/10.1007/978-981-16-8531-6_15

in an agriculture firm to obtain an exact value or a trend on a specify query "Carbon footprint per bale" from the vast amounts of related reports available online. Often the documents containing the term will be returned and then a manual search on the returned documents is required to identify the relevant value for the query of interest.

Information retrieval techniques have been applied in various scenarios to retrieve relevant information for users from a data collection based on user queries [3,4]. A user query is typically a short text with keywords representing the user requirement. Based on the given query, most relevant information is retrieved from the collection of data [5]. Majority of search engines including Google can only retrieve a list of documents that match the query, but they cannot return a value for the given query.

Corporations are under increasing pressure to produce annual reports describing their environmental, social, and economic sustainability [6]. These corporations set sustainability targets (e.g. reduce energy usage by 10%) and use data-derived indicators to measure progress towards those goals (e.g. total energy used in production). However, the data needed to generate these sustainability indicators are often stored in a number of disparate data sources [7]. To obtain domain-specific indicators, a manual search of information (i.e. values for indicators) from the data collection is the obvious choice. It is a cumbersome process to manually search for these indicator values using a search engine and then read through all relevant documents to identify the exact answer. An automated method of obtaining these domain-specific indicator values will be desirable.

With the recent advancements in automation, it is possible to semi automatically generate annual reports for companies that will contain information about data-derived indicators and targets set by the organization. This report will propose a system that could collect information on the indicators from various sources and, then synthesize the collected information to generate a report semi-automatically.

In this paper, we present a novel system that has been used in the Australian Cotton industry to help the users to sift through the Big data collected from various sources such as websites, published articles, case studies, reports and identify the value of the domain-specific indicators automatically. The methodology used in this system based on text mining approaches is used to scan and filter the relevant information from different unstructured data sources [2]. This information will then be collected to create a repository for inquiry and automatically generating a report with relevant values for the query of interest.

2   Background and Related Work

Content Extraction from Web Pages: The data in web pages are commonly represented as HTML and PDF documents where extraction of relevant content from these documents is essential to perform a text mining task [8]. Extraction of relevant content from HTML page is straightforward as the HTML documents are annotated with structured tags [9]. Mostly <p>, <a> and <table> tags cover the important information. While, it is easy to extract content in <p> and <a>, extracting meaningful information from tables is challenging. A heuristic approach can be used to extract table information in a cell format that makes the information retrieval easy [10]. Extracting content from PDF documents is challenging. The Optical Character Recognition (OCR) approach is widely used to read the contents of a PDF file in text format.

Chunking (Shallow parsing) and Named Entity Recognition (NER): Text mining techniques such as chunking and NER help to retrieve any phrase from the document that follows a regular expression and to identify the named entities in the document respectively. In chunking, a sentence or text is analysed by dividing it into syntactically related non-overlapping groups of words such as phrases [11]. NER is the task of identifying named entities (phrases that contains the names of persons, organizations or locations) in a text/sentence. Traditional approaches use sequential models like HMM (Hidden Markov Model) and CRF (Conditional Random fields) reply on handcrafted features [12]. Whereas the emerging approaches use neural network models such as RNN/LSTM with word embedding [13,14]. For a sentence that doesn't follow the given regular expression, Parts of speech (POS) can be used [15]. POS helps to focus only on the set of meaningful phrases in the documents and querying this set will return more relevant results. NER is necessary to identify the named entities in the document that can reveal the spatio-temporal information which is vital in many cases [16]. A simple spatio-temporal query needs spatio-temporal information to be known during the query and it is not possible to automatically reveal the spatio-temporal nature of the documents and hence a proper integration of Chunking and NER is essential to retrieve any results with underlying spatio-temporal information captured during the query process.

Indexing the Dataset: Indexing has been widely utilized to facilitate query searches [17]. Elasticsearch [18] is a popular document-oriented search engine that stores the data as an information repository. It provides the environment to deal with Big Data processing. Unlike traditional database systems, data is partitioned and stored as a collection of documents in Elasticsearch. It also maps the data using a dynamic mapping that makes it searchable using TF-IDF (Term-Frequency x Inverse Document Frequency) and BM25 methods [19,20].

Though each of these concepts has been used distinctively in various problems, the capability of these methods together is not well studied. In this paper, we integrate these concepts together to facilitate the framework for automated discovery of relevant information from the unstructured web data sources. We first extract contents from web pages and process various types of data. We perform chunking and NER on the pre-processed data that is subsequently stored and indexed in the Elasticsearch search engine. Finally, we perform query redefinition to fine tune to results.

3   Automated Discovery of Relevant Information

This section discusses about the five phases involved in the automated discovery of relevant information.

1. Data acquisition
2. Semantic enrichment
3. Data deposition
4. Query information
5. Visualization

We integrate text mining concepts to propose a novel methodology to automatically discover relevant focused information from the unstructured web data sources. As shown in Fig. 1, the proposed framework includes three main phases namely: data acquisition, semantic enrichment, and query formulation.

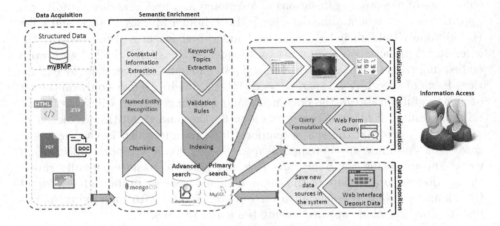

Fig. 1. The proposed framework

3.1   Data Acquisition

In this project, we use several types of data sources such as grey literature, web pages, and databases. Grey literature is often produced for internal communication or for public distribution by government agencies, professional organizations, universities, corporations, research centers, associations, and societies. This type of literature is not available through the usual bibliographic sources such as databases or indexes. Grey literature can be both in print and, increasingly due to digitization, it is in electronic formats as PDF, text and html pages. We envision using technical reports, census reports, survey reports, annual reports, journal articles and conference articles according to their availability and applicability.

In spite of the availability of several sources of data, regrettably, an increased quantity of data does not produce more value; only more of the relevant data will enable us to identify the indicator values correctly and autonomously. The next step is to clean the data for quality outcomes.

Data Deposition. Although this is one of the major steps, this is an often-neglected step. This step helps to improve the quality of the data that will eventually be used for knowledge discovery. Data gathering methods are often loosely controlled hence resulting in incorrect values and nearly impossible values too (e.g., age: 150 or male: yes and pregnant: true). Also, it is essential not to blindly remove noise as it can result in valuable information being unintentionally eliminated. It requires domain knowledge to remove the incorrect values with a correct definition of "noise". Otherwise, it can become cumbersome because pre-processing is reliant on the task at hand.

Text mining models require inputs to be presented in vector format. The transformation from the raw text data to a vector format requires several pre-processing steps such as parsing to remove unwanted tags and texts, stop words removal, stemming, term weighting generation and dimensionality reduction. We explain these steps below.

Data Parsing for Different Formats: Each of the data sources differs significantly in their structure and type of data that it holds. Hence, it is essential to deal with each of them using different techniques and identify relevant information that will be beneficial to users. The common data sources exist in the formats of HTML, Tables and PDFs. The following steps are required to convert these formats into text format so the querying process can begin to obtain the desired indicator value.

HTML Page: Web pages usually encode the data in a structured template that is suitable for human reading. It has to be converted in a format that is suitable for machines to process. Web scraping involves extracting text from structured documents like HTML, which uses tags that indicate the nature of data. For example, URLs are enclosed within <a> tag and paragraphs are enclosed within <p>. These tag structures are used to identify various elements within HTML documents. We identify all the elements containing text content and scrape them as a set of sentences in a text file.

Tables: The simple scraping of text from table elements (tags) is not suitable for processing as tables usually contain rich information and complex patterns with varying row and column representations. Tables often link their header rows and columns to cells that contain values representing header links as shown in Fig. 2. To extract meaningful information from a table, a machine learning algorithm called as Conditional Random Fields (CRF) has been applied. First it learns the column and row headers and then labels each line of the table to a header label. Using this approach, we achieve a more realistic data format of the table that is suitable for machines to process as shown in Fig. 3.

PDFs: Most of the data sources are reports and surveys, which are usually available as PDF files. Unlike HTML pages, PDF files do not have any structural information available as the metadata. This makes the analysis harder. We apply a PDF parsing technique that extracts only the text without any structural information. PDF parser extracts all the text represented as ASCII code. As PDFs are in human readable form, hence it does not have any structural

```
<QA_SECTION><QA_METADATA>
<TITLE> Onions: Area Planted and Harvested by Season State and United States 1996-98
</TITLE>
<CAPTIONS> Season Area Planted Area Harvested and State 1996 1997 1998 1996 1997 1998
Acres Spring </CAPTIONS>
<ROW> Spring AZ </ROW>
<COLUMN> Area Planted 1997 Acres </COLUMN>
</QA_METADATA>
2,100
</QA_SECTION
```

Fig. 2. Original table

Onions: Area Planted and Harvested by Season, State, and United States, 1996-98						
Season and State	: Area Planted		:	Area Harvested		
	: 1996 : 1997 : 1998		: 1996 : 1997 : 1998			
	:	Acres				
Spring	:					
AZ	: 2,100	2,100	2,500	1,900	2,100	2,500
CA	: 10,100	9,900	7,000	9,600	9,600	6,800
GA	: 16,000	16,200	15,000	14,700	15,800	13,900
TX	: 15,300	12,400	12,000	13,000	9,800	11,400
Total	: 43,500	40,600	36,500	39,200	37,300	34,600

Fig. 3. Extracted table headers

information for machine to perform tag based extraction. Hence text locations, font sizes and writing directions are used for focused parsing.

3.2   Semantic Enrichment Using Chunking and NER

The content, extracted from various sources like HTML pages, PDFs, and tables, is stored in a text file. We assume that the indicator values and associated information (i.e., respective value with units) can be identified correctly if chunking is done. Chunking is a process of identifying a group of chunks/information from a sentence based on Parts of Speech (POS). For each sentence, the words are tagged with their POS and Regular Expressions are used to identify the needed chunks. We assume that any indicators mentioned in a source have a high probability of being followed by its associated value and unit. The regular expressions are defined such that they can extract INDICATOR, VALUE and UNIT. For example, given the extracted text, "The average hectares planted per participant increased slightly 1,518 hectares.", the following information is identified:

1. Possible indicator: planted per participant
2. Possible value: 1518
3. Possible unit: hectares

The data sources contain various levels of spatial and temporal information. It is difficult for a simple querying process to identify them. We highlight the

spatial and temporal information from a source using NER. It generates anno-
tated tokens for entities in the text and highlights the names for each entity.
NER has near-human performance in identifying the following entities in the
English language, Location (Spatial), Organization, Date (Temporal), Money,
Person, Percent and Time (Temporal). An example is shown in Fig. 4 where
NER identifies two entities with spatial and temporal information. With the
annotated tokens for each sentence in a source, we can focus on obtaining only
the entities that we are interested.

The Australian cotton industry was the first agricultural industry in Australia to develop and document its
performance against specific environmental, economic and social sustainability indicators. Developed in
response to the industry's Third Environmental Assessment, the 2014 Australian Grown Cotton
Sustainability Report developed and benchmarked 45 key sustainability indicators for the Australian cotton
industry.

Potential tags:
LOCATION ORGANIZATION DATE MONEY PERSON PERCENT TIME

Fig. 4. NER example

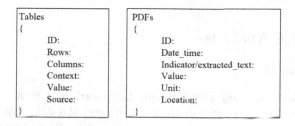

Fig. 5. The mapping process after parsing the information source

Indexing: Once all the data sources are collected, pre-processed and parsed, we
need to store the data and search the relevant information within the repository.
Indexing facilitates storing and allows a search engine to retrieve the data quickly.
The indexing will optimize and increase the runtime performance; thus finding
relevant information from a large data collection can be achieved within a few
milliseconds. As we have a distinct parsing technique for a Table source or a
PDF source, the data sources of Table and PDF formats are indexed with the
mapping process after parsing as shown in Fig. 5.

3.3   Query Formulation

With the information indexed, a query can be posed to the search engine based on
the mapping structure to retrieve the information about indicators. We start with

simple queries which directly searches for indicators without any modifications. For example, if the value of the indicator "area of cotton planted" is sought, then these four terms are formed as the query. Based on preliminary analyses, it was determined that the queries had to be modified or redefined in some cases to achieve satisfactory outcomes. We modified the query by adding more information. The additional information that found useful in extracting relevant results included

1. Simple Query
2. Simple Query + Keywords
3. Simple Query + Keywords + Source

Keywords. Keywords are usually terms that help to easily identify the indicators. We found that indicator unit made it easy to identify the indicator with high precision. Several keywords can be added to a simple query based on the domain knowledge.

Sources. Sometimes, even with the keywords provided, the query does not provide relevant results. If we know the data source from which the particular indicator can be identified, we can search the query for that particular source alone. By doing this one can avoid searching a big data source and thus any irrelevant information from unwanted sources can be ignored.

4 Empirical Analysis

4.1 Data Sources

The Cotton Research and Development Corporation (CRDC) is required to report multiple aspects of sustainability. However, the data sources where this information appears are de-centralised and are difficult to identify, and those that were identifiable were often appear in PDF format. This poses a significant challenge to CRDC to obtain relevant indicators values on repetitive basis. As a result, we recommend to use the proposed methodology to retrieve cotton-related data from a wide various sources in autonomous fashion.

There were a total of 102 sustainability indicators for which the information was required. Some example indicators inlcude: "Area of cotton planted"; "Irrigated planted area"; "Dryland planted area"; and "Gross value cotton lint". The possible data sources where this information can appear were obtained after consultation with domain experts. Some of the sources where the data was obtained are: Cotton Australia Website; Cotton Grower Yearbook 2016; Cotton Grower Surveys 2011, 2013 and 2018; Annual agricultural-commodities reports; Australia cotton shippers; and the Australian Bureau of Statistics (ABS) Website. There were several static data sources such as journal articles and technical reports.

4.2   Implementation

Open-source tools such as Python, MongoDB and Elasticsearch have been used to implement the proposed methodology. The unstructured data from various sources are parsed using Python and stored in a MongoDB database. The pre-processing, chunking and NER, as implemented in Python moduels, will decompose the extracted text from MongoDB into classes as shown in Fig. 5, which is then indexed in a local Elasticsearch server. The queries are fed into an Elasticsearch search engine to retrieve relevant indicator values. A simple user interface is built for user interaction.

4.3   Results: Information Retrieval Performance

Once the data sources were identified, we applied the proposed methodology to each indicator to obtain the results. We also assess how suitable and adaptable is a data source and the indicator for autonomous sustainability reporting purpose. Based on the categorization of each indicator, we identify if the particular indicator is adaptable.

Table 1 summarizes the findings for a few sample indicators and Table 2 details the overall results. A total of 102 indicators deemed to retrieve high performance (i.e., high relevance score). A total of 15 indicators had moderate values; however, we were unable to locate data for 23 indicators. The reason may be that data may not exist for these indicators and need specific help to collect the possible data sources where the values may reside. We plan to implement some sophisticated machine learning methods to identify values of these indicators if they exist in collected sources. We also plan to have workshops with CRDC staff to get more information on data sources where this information may appear.

For every indicator, we analyze the query results in depth to identify Suitability and Adaptability of the current sources in the repository.

Suitability: We measure the suitability to identify if the existing data is suitable for the development of the Sustainability report. Based on the query analysis, we categorize suitability as High, Medium and Low.

Each query returns a list of matching results. Each returned result is ranked to identify the best match. We use the normalized relevance score to decide the suitability category of the indicators. The normalized relevance score is to measure how relevant is the results retrieved for the given query. The relevance score is calculated using term frequency $\times$ inverse document frequency, and field length norm. These measures are combined and normalized for each source type to rank the indicators. Higher the value, the higher is the suitability (Table 3).

Adaptability: We use adaptability to identify the continuous (ongoing) extraction capacity of the indicators and the sources. We categorize adaptability using two factors as shown in Table 4. The adaptability score is an integrated score derived from two scores related to query and data dependencies as shown in Table 5.

Table 1. Result summary

S. No	Indicator	Query	Data source	Source type	Added keywords	Suitability	Adaptability	Relevance score	Result achieved
1	Area of cotton planted	Cotton area planted 2016	D1	Table	ha	H	H	0.59	Y
2	Irrigated planted area	Irrigated planted area	D2	PDF	ha	M	L	0.30	Y
3	Cotton exports	Export million tonnes	D3	PDF	Million tonnes	M	M	0.40	Y
4	Total amount of cotton produced	Total amount of cotton produced (metric tonnes OR million bales)	D1	HTML	Metric tonnes OR million bales	H	H	0.73	Y
5	Percentage change: Crop rotation	Crop rotation	D4	PDF	%	L	M	0.63	Y
6	Native vegetation area per farm	Native vegetation area per farm	D4	HTML	Percentage farm land	M	H	0.48	Y
7	Regional gross production value	Regional gross production value	D3	PDF	%	M	M	0.28	Relevant results
8	Average cotton area per farm	Average cotton area per farm	D1	Table	ha/farm	M	H	0.22	Y
9	Salinity	Salinity	D4	PDF	mm/year	L	L	0.00	N
10	Percentage of growers using integrated pest management	Percentage of growers using integrated pest management	D5	PDF	%	H	H	0.63	Y
11	Irrigation water use index	Water use on Australian farms cotton	D6	HTML	%	M	H	0.24	Y
12	Groundwater levels	Groundwater levels	D7	PDF	Million tonnes	H	H	0.69	Y
13	Workers receiving regular health and safety training	Workers receiving regular health and safety training	D4	PDF	% respondents or %	M	M	0.35	Y

Table 2. Overall results

Data Type	Total Queries	Results achieved	Relevant results	Results not achieved
HTML	13	10	3	0
PDF	76	51	12	13
Table	3	3	0	0
Unknown	10	0	0	10
Total	102	64	15	23

Table 3. Suitability score

Rank level	Score
High	0.7 to 1
Medium	0.4 to 0.6
Low	0 to 0.3

Table 4. Adaptability categories

Rank level	Query-dependent	Data-dependent
High	2 or more query re-definitions	Source subscription and source-specific search
Medium	1 query re-definition	Source specific search
Low	No query redefinition	Open source

Table 5. Adaptability score

Query-dependent	Data-dependent	Adaptability score
L/M	L	H
L/M	M	M
L/M	H	L
H/M	L	M
H/M	M/H	L

Query-dependent: Any indicator that is query dependent is considered as medium/high category. Query-dependent is a condition when a simple query with keywords does not result in a good retrieval. Sometimes, the indicators are represented in different terms in the sustainability report and the data source. The situation is similar when the source uses different units to measure the same indicator. By carefully redefining the query with the possible alternative terms

Table 6. Query redefinition

Indicator	Simple query	Keyword (Query dependent)	Source (Data dependent)	Result achieved	Relevance score
Irrigated planted area	Irrigated planted area			Y	13.35
Cotton exports	Cotton exports			N	9.67
Cotton exports	Cotton exports	Million tonnes		Y	**17.57**
Cotton stubble	Cotton stubble			N	12.34
Cotton stubble	Cotton stubble	%		N	12.34
Cotton stubble	Cotton stubble	%	D4	Y	**17.03**

and units, this can be avoided to get better query results. Out of 102 indicators, 66 indicators are retrieved with high performance using units as keyword whereas only 48 indicators can be retrieved without keywords. We categorize the query-dependent as high, medium and low based on the level of query redefinition required (Table 5).

Data-dependent: Any indicator that is data dependent is considered as a medium/high category. Data-dependence is a condition when you need a subscription to access the source data or when you do a source-specific search. Even though source specific searches will not affect in retrieving relevant results, in future if that specific source is not available, the indicator may be difficult to query. Table 6 shows an example of 3 indicators where the importance of Keywords and Data source in retrieving relevant results are analyzed. While "Irrigated planted area" can return the required result without any query redefinition, "Cotton exports" needs "Million tonnes" as a keyword to achieve the result. On the other hand, indicator "Cotton stubble" is data dependent and results may not be achieved if that specific data source (D4) is not available in the repository.

5   Conclusion

With the growing advances in the web and the collection of unstructured data, it is crucial to process and analyze them for various purpose. Especially when the data sources have textual information in the different forms, the retrieval of relevant information becomes challenging. In this paper, we propose a methodology to automatically discover the relevant information from the collection of unstructured web data sources. We combine several text mining techniques to extract, store and perform a query search.

   We implemented this methodology for a corporation for their specific need of obtaining information on sustainability indicators autonomously. We presented an overall results as well as we provided a qualitative analysis detailing the suit- ability and adaptability of the query and data sources respectively. The proposed method enables the corporation to utilise the Information Repository built from a wide variety of sources for interrogating the relevant information for sustainability reporting.

The proposed method can be applied to similar problems. In future, we will adopt a learning-to-rank concept in query redefinition that eases the human input.

References

1. Krallinger, M., Rabal, O., Lourenco, A., Oyarzabal, J., Valencia, A.: Information retrieval and text mining technologies for chemistry. Chem. Rev. **117**(12), 7673–7761 (2017)
2. Banawan, K., Ulukus, S.: The capacity of private information retrieval from coded databases. IEEE Trans. Inf. Theory **64**(3), 1945–1956 (2018)
3. Croft, W.B., Metzler, D., Strohman, T.: Search Engines: Information Retrieval in Practice, vol. 520, Addison-Wesley Reading, Boston (2010)
4. Mayer-Schönberger, V., Cukier, K.: Big data: A revolution that will Transform how we Live, Work, and Think. Houghton Mifflin Harcourt, Boston (2013)
5. Wu, X., Zhu, X., Wu, G.Q., Ding, W.: Data mining with big data. IEEE Trans. Knowl. Data Eng. **26**(1), 97–107 (2014)
6. Porter, M.E., Kramer, M.R.: The link between competitive advantage and corporate social responsibility. Harvard Bus. Rev. **84**(12), 78–92 (2006)
7. Peterson, E.E., Cunningham, S.A., Thomas, M., Collings, S., Bonnett, G.D., Harch, B.: An assessment framework for measuring agroecosystem health. Ecol. Ind. **79**, 265–275 (2017)
8. Gupta, S., Kaiser, G., Neistadt, D., Grimm, P.: Dom-based content extraction of html documents. In: Proceedings of the 12th International Conference on World Wide Web, pp. 207–214. ACM (2003)
9. Lin, S.H., Ho, J.M.: Discovering informative content blocks from web documents. In: Proceedings of the Eighth ACM SIGKDD International Conference on Knowledge Discovery and Data Mining, pp. 588–593. ACM (2002)
10. Wei, X., Croft, B., McCallum, A.: Table extraction for answer retrieval. Inf. Retrieval **9**(5), 589–611 (2006)
11. Sang, E.F., De Meulder, F.: Introduction to the conll-2003 shared task: Language-independent named entity recognition. arXiv preprint cs/0306050 (2003)
12. Finkel, J.R., Grenager, T., Manning, C.: Incorporating non-local information into information extraction systems by Gibbs sampling. In: Proceedings of the 43rd Annual Meeting on Association for Computational Linguistics, pp. 363–370. Association for Computational Linguistics (2005)
13. Zhai, F., Potdar, S., Xiang, B., Zhou, B.: Neural models for sequence chunking. In: Thirty-First AAAI Conference on Artificial Intelligence (2017)
14. Habibi, M., Weber, L., Neves, M., Wiegandt, D.L., Leser, U.: Deep learning with word embeddings improves biomedical named entity recognition. Bioinformatics **33**(14), i37–i48 (2017)
15. Grishman, R.: Information extraction: techniques and challenges. In: Pazienza, M.T. (ed.) SCIE 1997. LNCS, vol. 1299, pp. 10–27. Springer, Heidelberg (1997). https://doi.org/10.1007/3-540-63438-X_2
16. Strötgen, J., Gertz, M., Popov, P.: Extraction and exploration of spatio-temporal information in documents. In: Proceedings of the 6th Workshop on Geographic Information Retrieval, p. 16. ACM (2010)

17. Kononenko, O., Baysal, O., Holmes, R., Godfrey, M.W.: Mining modern repositories with elasticsearch. In: Proceedings of the 11th Working Conference on Mining Software Repositories, pp. 328–331. ACM (2014)
18. Akdogan, H.: Elasticsearch Indexing. Packt Publishing Ltd, Birmingham (2015)
19. Ramos, J., et al.: Using tf-idf to determine word relevance in document queries. In: Proceedings of the First Instructional Conference on Machine Learning, vol. 242, pp. 133–142 (2003)
20. Pérez-Iglesias, J., Pérez-Agüera, J.R., Fresno, V., Feinstein, Y.Z.: Integrating the probabilistic models bm25/bm25f into lucene. arXiv preprint arXiv:0911.5046 (2009)

Nonnegative Matrix Factorization to Understand Spatio-Temporal Traffic Pattern Variations During COVID-19: A Case Study

Anandkumar Balasubramaniam[1], Thirunavukarasu Balasubramaniam[2],
Rathinaraja Jeyaraj[1], Anand Paul[1(✉)], and Richi Nayak[2]

[1] School of Computer Science and Engineering, Kyungpook National University,
Daegu, South Korea
{bsanandkumar,jrathinaraja,anand}@knu.ac.kr
[2] School of Computer Science and Centre for Data Science,
Queensland University of Technology, Brisbane, Australia
{thirunavukarasu.balas,r.nayak}@qut.edu.au

Abstract. Due to the rapid developments in Intelligent Transportation System (ITS) and increasing trend in the number of vehicles on road, abundant of road traffic data is generated and available. Understanding spatio-temporal traffic patterns from this data is crucial and has been effectively helping in traffic plannings, road constructions, etc. However, understanding traffic patterns during COVID-19 pandemic is quite challenging and important as there is a huge difference in-terms of people's and vehicle's travel behavioural patterns. In this paper, a case study is conducted to understand the variations in spatio-temporal traffic patterns during COVID-19. We apply nonnegative matrix factorization (NMF) to elicit patterns. The NMF model outputs are analysed based on the spatio-temporal pattern behaviours observed during the year 2019 and 2020, which is before pandemic and during pandemic situations respectively, in Great Britain. The outputs of the analysed spatio-temporal traffic pattern variation behaviours will be useful in the fields of traffic management in Intelligent Transportation System and management in various stages of pandemic or unavoidable scenarios in-relation to road traffic.

Keywords: Traffic pattern · NMF · Pattern mining · Spatio-temporal analysis · COVID-19

This research was supported by the National Research Foundation of Korea (Grant No. 2020R1A2C1012196), and in part by the School of Computer Science and Engineering, Ministry of Education, Kyungpook National University, South Korea, through the BK21 Four Project, AI-Driven Convergence Software Education Research Program, under Grant 4199990214394.

1   Introduction

Vehicular traffic pattern analysis is an important topic in developing/improving traffic management system, especially in ITS [1]. Due to the latest developments and technological advancements in ITS, the availability of the spatio-temporal traffic data is abundant [2,3]. However, the traffic patterns are heterogeneous in such a way that various road segments have distinct time-varying traffic patterns [4]. Addressing this problem of heterogeneity by analysing various features such as location, time, etc., is much needed to get useful traffic pattern insights from this abundance of spatio-temporal traffic data. Also, there are various factors such as congestion, accidents, natural calamities, weather conditions, etc., that affect these vehicular traffic patterns in different ways.

Having said that, the year 2020 seemed to have a lot of uncertainties in people's travel behaviour due to the rapid increase in COVID-19 cases [5,9]. Due to which, people around the world have faced unimaginable disturbances and struggles that had adversely affected the usual travel patterns of people and vehicles due to the additional enforcement of various stages of curfew. In this paper, vehicular traffic data recorded before and during the pandemic in Great Britain is analyzed; and discussed how the spatio-temporal patterns are getting varied due to the COVID-19 pandemic [6] in 2020 in Great Britain.

To understand the spatio-temporal traffic pattern variations due to COVID-19, in this paper, we use NMF [7]. We apply NMF on traffic data of Grate Britain generated during 2019 and 2020 to have a meaningful insights of spatio-temporal traffic patterns [8] before and during COVID-19 pandemic i.e. in the year 2019 and 2020 respectively. The case-study provided in this paper helps to identify pattern variations and to understand the impacts of travel or traffic patterns due to COVID-19. While the simple vehicle counts will provide knowledge on how much traffic pattern has changed, it cannot say what patterns have changed. This paper specifically focuses on learning what and how patterns have changed during COVID-19.

The rest of this paper is organized as follows: Sect. 2 contains related works, Sect. 3 includes the background on NMF, and discuss the traffic pattern elicitation process and the importance of understanding the traffic pattern variations. Section 4 contains the case study that is conducted on the 2019 and 2020 traffic dataset of Great Britain, which includes dataset descriptions, evaluation measures to run the NMF model followed by spatio-temporal traffic pattern analysis results of 2019 and 2020. Section 5 concludes the paper and briefly introduces future directions.

2   Related Work

Pattern mining is a subset of data mining, which helps to discover or analyse various latent features or insights from high dimensional data. It is commonly applied on vehicular traffic data to analyse the spatio-temporal vehicular traffic patterns [11,12]. For example, in [11], authors dealt with the problem of identifying vehicular traffic patterns and congestion problems occurring in densely populated cities. The authors in [11] came up with the optimal traffic solution using social media data records. Historical data about social media posts and tweets are taken into consideration and social media knowledge related to the traffic is analysed to predict the traffic pattern at a given location. However, this might not be suitable during pandemic situations like COVID-19 as the mobility pattern of the people changes drastically. In another work [12], authors dealt with the traffic forecasting problem by implementing Fourier analysis with wavelet denoising technique along with two layer Fast Fourier Transform to recognize and formulate the periodic pattern. However, the approach is not capable to elicit spatial patterns.

Similarly, many works have focused on learning traffic patterns to perform different downstream tasks during COVID-19. For instance authors in [13] dealt with traffic congestion prevention by learning traffic trends using linear regression. Authors in [14] utilized the crowd-sourced traffic congestion data to determine mobility changes in Manhattan, NYC during COVID-19. Time-series decomposition method is used to quantify the changes in various traffic categories. Crowd-sourced traffic data is utilized to inform human mobility changes to plan for the future pandemic responses. The usage of crowd-sourced traffic data might not be reliable due to authenticity of data. Authors in paper [15] proposed morphology-based vehicle detection method to detect traffic density in multiple cities and compare traffic data with mobility pattern changes due to COVID-19. Morphology-based vehicle detection technique requires the usage of utilizing image-based dataset, which might not be feasible every-time due to the unavailability of image data. In [16], by using high-resolution remote sensing imagery, traffic density reduction pattern across the various parts of the world during COVID-19 is analysed. Though the results are helpful, the association between temporal and spatial patterns is not learned in their work.

Recently, factorization-based techniques have been used to learn the associations between spatial and temporal patterns. For instance, authors in [17] used NMF and Nonnegative Tensor Factorization to learn the spatio-temporal dynamics of topics in twitter during COVID-19. In this paper, we investigate the usage of NMF to learn spatio-temporal patterns from traffic data and understand the pattern variations during COVID-19.

In this paper, the challenges of combining or comparing the temporal and spatial patterns are addressed in such a way that both the temporal and spatial patterns of the recorded traffic data in Great Britain during the year 2019 and 2020 are analysed and compared along with the utilization of NMF model to understand the spatio-temporal pattern variations before and during COVID-19 pandemic.

3 Nonnegative Matrix Factorization for Traffic Pattern Elicitation

3.1 Nonnegative Matrix Factorization

NMF is one of the popular dimensionality reduction techniques which uses nonnegativity constraint to generate non-negative lower-dimensional representations of a matrix, that further lead to the increase in the quality of matrix factorization. Suppose $\mathbf{X} \in \mathbb{R}^{(N \times M)}$ and $\mathbf{Y} \in \mathbb{R}^{(A \times B)}$ are two matrices, they can be factorized into two lower-dimensional factor matrices $U \in \mathbb{R}^{(N \times r)}$ and $V \in \mathbb{R}^{(M \times r)}$ & $P \in \mathbb{R}^{(A \times k)}$ and $Q \in \mathbb{R}^{(B \times k)}$, respectively. The factorization of matrix can be defined as,

$$\mathbf{X} \approx UV^T \quad s.t, U \geq 0 \quad \text{and} \quad V \geq 0, \tag{1}$$

$$\mathbf{Y} \approx PQ^T \quad s.t, P \geq 0 \quad \text{and} \quad Q \geq 0. \tag{2}$$

3.2 NMF for Traffic Pattern Elicitation

In traffic scenario, location-time matrices $\mathbf{X} \in \mathbb{R}^{(N \times M)}$ (which stores the 2019 traffic record) and $\mathbf{Y} \in \mathbb{R}^{(A \times B)}$ (which stores the 2020 traffic record) are the representation of the count of vehicles with respect to given location and time. N, A in is the total number of unique locations i.e. 12656 locations in 2019, 6537 locations in 2020; and M, B is the range of time records i.e. 12 hours from 07:00 to 18:00 h with the interval of 1 h.

NMF is very useful in separately extracting the spatio-temporal patterns by representing (location × features) and (time × features) individually for 2019 and 2020, in which each column of the factor matrices is a pattern or feature. Based on the rank r defined in Eq. 1 and k in Eq. 2, r and k distinctive patterns can be obtained from $\mathbf{X}$ and $\mathbf{Y}$ during the factorization process respectively.

From Eq. 3 & 4, the optimization problem is formulated as a minimization problem with Euclidean distance as the cost function individually for 2019 and 2020 as below,

$$\min_{U \geq 0, V \geq 0} f(U, V) = \left\| \mathbf{X} - UV^T \right\|_2 \quad s.t, U \geq 0 \quad \text{and} \quad V \geq 0 \tag{3}$$

$$\min_{P \geq 0, Q \geq 0} f(P, Q) = \left\| \mathbf{Y} - PQ^T \right\|_2 \quad s.t, P \geq 0 \quad \text{and} \quad Q \geq 0 \tag{4}$$

Determination of ranks to run the NMF model on both of the generated location-time matrices is discussed elaborately in Sect. 4.2. Nonnegative lower-dimensional representation of the matrices $\mathbf{X}$ and $\mathbf{Y}$ generated on performing the NMF model contains (location × patterns) and (time × patterns) as U and V for $\mathbf{X}$, P and Q for $\mathbf{Y}$. These NMF-generated location and time patterns are grouped into individual location clusters and time clusters, respectively.

3.3 Understanding Traffic Pattern Variations

Spatio-temporal traffic patterns in ITS are much essential for better understanding the latent spatio-temporal variation patterns. These observations of pattern variations help us to predict or manage the traffic scenario prior to various application based use-cases and help to improve the traffic management across the country which further leads to the development of smart environment in ITS.

The temporal variation patterns observed in 2019 are compared with the corresponding spatial variation patterns as well as compared to the temporal variation patterns of 2020. This pattern comparison is very useful in unexpected scenarios such as COVID-19 pandemic, and helps to get various insights of the vehicular traffic pattern distributions.

4   A Case Study

4.1   Dataset

The case study is performed on the dataset [10], which is downloaded from the repository of the Department of Transport, Great Britain. Dataset contains the count of vehicles that is recorded in major and minor roads of Great Britain. The data is filtered to include traffic records for the year 2019 and 2020 only. Figure 1 shows the traffic spread of the count of various kinds of passing vehicles that were recorded from 0700 to 1800 hours during the year 2019 and 2020 across Great Britain. This figure explicitly shows that there is a drastic decrease in the total number of vehicles on roads. However, this could not answer how patterns are varied during COVID-19.

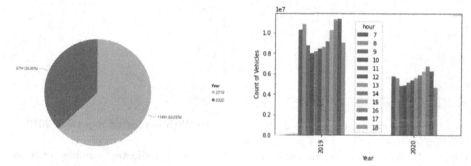

(a) Spread of count of vehicles in Great Britain during 2019 and 2020

(b) Hourly comparison of vehicle counts in the year 2019 and 2020

Fig. 1. Spread of vehicle counts in Great Britain

To generate location-time matrices $\mathbf{X}$ and $\mathbf{Y}$ corresponding to the years 2019 and 2020, respectively, 12656 unique locations with 12 h time from 2019 and 6537 unique locations with 12 h time from 2020 are grouped according to the location & hour and the cumulative sum of the total number of recorded motor vehicles at the given location and time is calculated such that the input matrices $\mathbf{X}$ and $\mathbf{Y}$ are in the order of (12656×12) and $((6537 \times 12))$, respectively.

The generated matrices contain numerical values, which need to be normalized in prior to the NMF modeling to make NMF perform better on the generated location-time matrices. The process of performing Normalization on the input matrices scales the numerical variables into nonnegative floating-point variables of range 0 to 1. Data Normalization is achieved by using MinMax Scalar Transforms. The normalized matrix is further analyzed to determine the rank using various evaluation measures such as Calinski and Harabasz score which is the ratio between with-in cluster[1] and between-cluster dispersion.

4.2   Evaluation Measures

In follow-up towards the step of Normalization, evaluation measures are performed mainly to determine the rank that needs to be given as an input to the NMF model to generate temporal and spatial patterns for 2019 and 2020. To determine the optimal rank, with-in cluster dispersion (Eq. 5) and between-cluster dispersion (Eq. 6) analysis is carried out on the clusters and datapoints. The value should be low for with-in cluster dispersion evaluation and high for between-cluster dispersion evaluation which in-turns proves that the clusters should be well separated from other clusters and the datapoints within cluster should be grouped together, respectively.

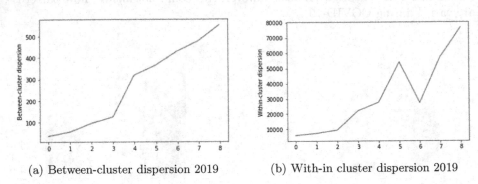

(a) Between-cluster dispersion 2019   (b) With-in cluster dispersion 2019

Fig. 2. Rank evaluation measures for 2019. (x-axis indicates number of clusters/patterns.

$$W_d = \sum_{g=1}^{m} \sum_{x \epsilon C_g} (x - c_g)(x - c_g)^T \qquad (5)$$

$$B_d = \sum_{g=1}^{m} n_g(c_g - c_S)(c_g - c_S)^T \qquad (6)$$

where B_d is between-cluster dispersion, W_d is within-cluster dispersion, S indicates the set of data, g is clusters, C_g is the set of points in the cluster, c_g is the center of the cluster, c_S is the center of data, and n_g is the number of points in the cluster.

---

[1] Cluster is nothing but pattern.

Figure 2 shows the with-in cluster dispersion and between-cluster dispersion evaluation for the year 2019. From the Fig. 2a, it is clear that there is an increase in the trend of between-cluster dispersion graph values 3, 4 and 6. In addition to this, from Fig. 2b, the corresponding values 3, 4 and 6 are analysed and it is noticeable that within-cluster value of 6 is quite convincing as the value trend should be opposite to that of the between-cluster value trend. In this scenario, there is a hike in the value 6 of Fig. 2a and corresponding drop in the value 6 of Fig. 2b. From this analysis, one can conclude that the association of data points is good at the value 6, hence the rank value is set to be as 6 for running the NMF model in 2019.

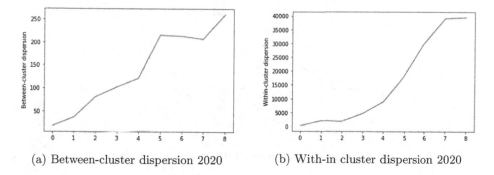

(a) Between-cluster dispersion 2020   (b) With-in cluster dispersion 2020

Fig. 3. Rank evaluation measures for 2020

For the 2020 datapoints, with-in cluster dispersion and between-cluster dispersion are carried out and the evaluation measures are recorded in Fig. 3. The between-cluster dispersion values are varying across the values 2, 4, 5 and 7 and corresponding with in cluster dispersion values are to be taken into consideration in determining the rank to perform NMF model on the 2020 traffic datapoints. From the evaluation, it is obvious to take the rank value as 4 to compare the relationship of the spatio-temporal patterns. This already hints that in 2020 during COVID-19, two spatio-temporal patterns have disappeared. Further analysis is needed to find out what that pattern is?

4.3 Traffic Pattern Results for 2019

Figure 4 shows the traffic patterns obtained from the lower-dimensional factor matrix V on performing NMF model in normalized **X** matrix belonging to 2019 traffic records. According to the rank obtained from evaluation measures, there are 6 temporal patterns (p1 to p6) observed. The x-axis in Fig. 4 indicates record from 07 : 00 to 18 : 00 hours, and y-axis indicates the intensity of vehicles at the given time. On analysing the observed patterns, it is very clear that pattern (p6) records high traffic data in the early morning, pattern (p5) records high traffic data in the evening followed by pattern (p3) which records the second highest traffic count in the evening. The pattern (p2) shows the decrease of traffic density and moderate increase of the traffic density in morning and evening, respectively. Comparatively, the pattern (p1) shows very low traffic density record during evening peak times. In contrast to all, there is an abnormal traffic behaviour pattern in (p4) during the period 10 : 00 to 14 : 00 hours in comparison to all other pattern behaviours.

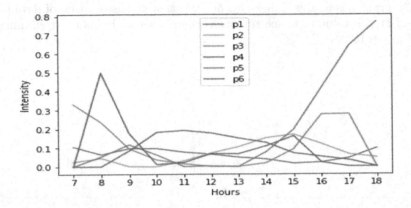

Fig. 4. Temporal patterns in 2019. (Color figure online)

These pattern visualizations show the different temporal behaviours of people's mobility. The insights from Fig. 4 are: the number of people who are active in the morning are not very much active in the evening, and vice-versa. This is justified with the insights from p6 and p2, respectively. The corresponding spatial behavioural patterns in 2019 are obtained from the lower-dimensional location factor matrix U on performing NMF model in the normalized $\mathbf{X}$ matrix in relation to the above temporal patterns. In terms of spatial relationship, there is only a minor difference between the spatial patterns as observed in Fig. 5. However, if we look into spatial pattern (purple colour) in comparison with the temporal pattern (p5) from Fig. 4, there is a difference such that the pattern (p5) shows the peak in the evening but in parallel the spread of spatial vehicle density in corresponding to the pattern (p5) is less in comparison to that of the other patterns from Fig. 5.

4.4   Traffic Pattern Results for 2020

Temporal pattern observations for 2020 with 4 clusters is plotted in Fig. 6, which shows the temporal pattern elicited from the lower-dimensional representation matrix Q of the $\mathbf{Y}$ matrix by running the NMF model with rank 4. Figure 6 shows the temporal patterns observed during 07 : 00 to 18 : 00 h during the COVID-19 pandemic year 2020, and it is very obvious that two patterns are already missing in 2020 in-comparison to the temporal patterns observed in 2019.

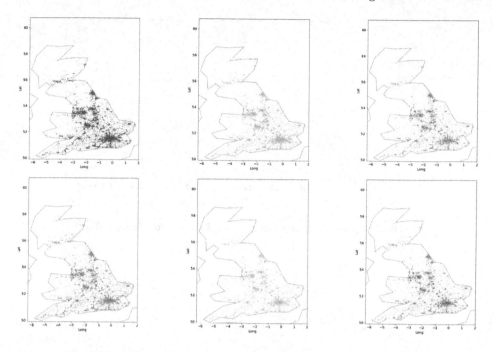

Fig. 5. Spatial patterns in 2019 with 6 clusters (Color figure online)

Figure 6 shows the temporal patterns observed with 4 clusters in which the pattern (p3) records high traffic density in the morning with a gradual decrease in the afternoon period and sudden increase of traffic density in the evening peak hours. However, the overall traffic density is 52% less than the year 2019. Though the other patterns (p1, p2, p4) have a gradual increase and decrease during the overall day, the pattern (p3) is dominating in the morning and evening, respectively.

The corresponding spatial behavioural patterns in 2020, shown in Fig. 7, obtained from the lower-dimensional location factor matrix P on performing NMF model in the normalized **Y** matrix are shown in relation to the temporal patterns in Fig. 6. According to the spatial pattern observations, though the pattern (blue) is diversely spread across the country in comparison to the other patterns, the temporal behaviours of the corresponding pattern (p1) in Fig. 6 are moderate in comparison to the other temporal pattern behaviours.

4.5   Traffic Pattern Variation Analysis

The importance of understanding the traffic pattern variations is much essential for building a smarter and efficient transportation system. On comparing the temporal patterns of before COVID-19 and during COVID-19 scenarios, it is very obvious that the vehicular mobility patterns of the people who used to travel in the early morning are drastically/moderately changed in the corresponding

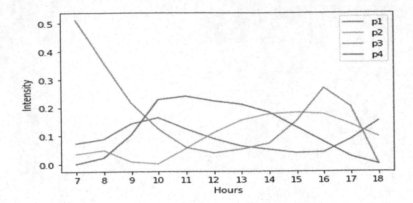

Fig. 6. Temporal variation patterns in 2020 (Color figure online)

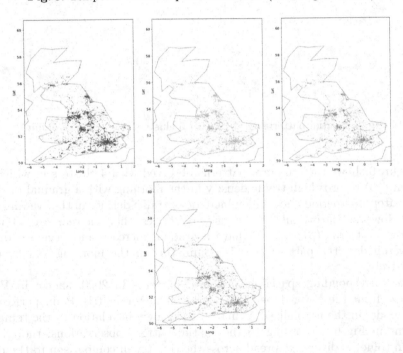

Fig. 7. Spatial patterns in 2020 with 4 clusters (Color figure online)

patterns in 2020, and there is a disappearance of two patterns when compared to 2019 temporal behaviours. In particular, vehicular traffic pattern (p6) is higher in the morning and slight increase in the evening with a fluctuation in-between. However, the corresponding pattern is completely not present in 2020 due to the sudden influence of COVID-19 cases.

On the other hand, the temporal patterns (p3 and p4) in 2019 and 2020 show variations in behaviour accordingly. These temporal variations in certain patterns show that the people who are active in the morning were reduced and vice-versa, and to be noticed that the people's activity is increased in couple of the patterns of 2020. However, in comparison, the pattern (p1) in 2019 is almost similar to the pattern (p1) in 2020. In parallel, comparison to the spatial patterns observed in 2019 and 2020, there is a very small difference in spatial pattern variations however there is a spatio-temporal pattern variation observed within the years 2019 and 2020. The effect of pandemic plays a major role in the temporal patterns of Great Britain with only few spatial pattern density differences observed for 2019 and 2020.

5 Conclusion

This paper analyses the spatio-temporal traffic variation patterns recorded between 0700 to 1800 hours during the year 2019 (before-COVID-19) and 2020 (during-COVID-19) in Great Britain. We also demonstrate the capability of utilizing NMF in getting various useful insights from the traffic patterns elicited from the NMF outputs. The case study conducted in this paper provides detailed traffic patterns and shows the variations in the spatio-temporal patterns. Our observations from the temporal and spatial patterns show that the temporal traffic patterns are drastically varied during COVID-19, while there is only a few minor variations of spatial traffic patterns. This spatio-temporal traffic pattern variation insight will lead to more understanding about the abnormal behaviour of the traffic mobility patterns compared to the past historical data and further leads to improvising the traffic management planning in the future pandemic or unavoidable scenarios. In future, a more detailed analysis will be conducted to see how patterns change over the period of month or week.

Acknowledgement. This research was supported by the National Research Foundation of Korea (Grant No. 2020R1A2C1012196), and in part by the School of Computer Science and Engineering, Ministry of Education, Kyungpook National University, South Korea, through the BK21 Four Project, AI-Driven Convergence Software Education Research Program, under Grant 4199990214394.

References

1. Lin, Y., Wang, P., Ma, M.: Intelligent transportation system (ITS): concept, challenge and opportunity. In: IEEE 3rd International Conference on Bigdata Security on Cloud, IEEE International Conference on High Performance and Smart Computing (HPSC), and IEEE International Conference on Intelligent Data and Security (IDS), pp. 167–172. IEEE (2017)
2. Xia, D., et al.: Discovering spatiotemporal characteristics of passenger travel with mobile trajectory big data. Physica A Stat. Mech. Appl. **578**, 126056 (2021)
3. Wang, S., Cao, J., Yu, P.: Deep learning for spatio-temporal data mining: a survey. IEEE Trans. Knowl. Data Eng. (2021). https://doi.org/10.1109/TKDE.2020.3025580

4. He, P., Jiang, G., Lam, S.-K., Sun, Y.: Learning heterogeneous traffic patterns for travel time prediction of bus journeys. Inf. Sci. **512**, 1394–1406 (2020)
5. Ozili, P.K., Arun, T.: Spillover of COVID-19: impact on the global economy. SSRN (2020)
6. United Kingdom Coronavirus Map and Case Count - The New York Times. https://www.nytimes.com/interactive/2021/world/united-kingdom-covid-cases.html, Accessed 07 Sept 2021
7. Vendrow, J., Haddock, J., Rebrova, E., Needell, D.: On a guided nonnegative matrix factorization. In: IEEE International Conference on Acoustics, Speech and Signal Processing, pp. 3265–32369. IEEE (2021)
8. Balasubramaniam, T., Nayak, R., Bashar, M.A.: Understanding the spatio-temporal topic dynamics of covid-19 using nonnegative tensor factorization: a case study. In IEEE Symposium Series on Computational Intelligence (SSCI), pp. 1218–1225. IEEE (2020)
9. Saeed, F., Paul, A., Ahmed, M.J.: Forecasting COVID-19 cases using multiple statistical models. In 8th International Conference on Orange Technology (ICOT), pp. 1–5. IEEE (2020)
10. Vehicle counts recorded on major and minor roads. https://roadtraffic.dft.gov.uk/#6/55.254/-6.053/basemap-regions-countpoints, Accessed 07 Sept 2021
11. Shekhar, H., Setty, S., Mudenagudi, U.: Vehicular traffic analysis from social media data. In: International Conference on Advances in Computing, Communications and Informatics (ICACCI), pp. 1628–1634. IEEE (2016)
12. Sun, P., AlJeri, N., Boukerche, A.: A fast vehicular traffic flow prediction scheme based on fourier and wavelet analysis. In: IEEE Global Communications Conference (GLOBECOM), pp. 1–6. IEEE (2018)
13. Sinha, A., Puri, R., Balyan, U., Gupta, R., Verma, A.: Sustainable time series model for vehicular traffic trends prediction in metropolitan network. In: 6th International Conference on Signal Processing and Communication (ICSC), pp. 74–79 (2020)
14. Jenni, A., Shearston, M.E., Martinez, Y.N., Markus, H.: Social-distancing fatigue: evidence from real-time crowd-sourced traffic data. Sci. Total Environ. **792**, 148336 (2021)
15. Chen, Y., Qin, R., Zhang, G., Albanwan, H.: Spatial temporal analysis of traffic patterns during the COVID-19 epidemic by vehicle detection using planet remote-sensing satellite images. Remote Sens. **13**, 208 (2021)
16. Wu, C., et al.: Traffic density reduction caused by city lockdowns across the world during the COVID-19 epidemic: from the view of high-resolution remote sensing imagery. IEEE J. Sel. Topics Appl. Earth Obser. Remote Sens. **14**, 5180–5193 (2021)
17. Balasubramaniam, T., Nayak, R., Luong, K., Bashar, M.A.: Identifying Covid-19 misinformation tweets and learning their spatio-temporal topic dynamics using nonnegative coupled matrix tensor factorization. Social Netw. Anal. Mining **11**(1), 1–19 (2021). https://doi.org/10.1007/s13278-021-00767-7

Correction to: Taking the Confusion Out of Multinomial Confusion Matrices and Imbalanced Classes

David Lovell[ID], Bridget McCarron[ID], Brendan Langfield,
Khoa Tran[ID], and Andrew P. Bradley[ID]

Correction to:
Chapter "Taking the Confusion Out of Multinomial
Confusion Matrices and Imbalanced Classes"
in: Y. Xu et al. (Eds.): ***Data Mining*, CCIS 1504,**
https://doi.org/10.1007/978-981-16-8531-6_2

In the originally published version of chapter 2, the Table 1. contained an error in a formula. The formula error in Table 1. has been corrected.

The updated version of this chapter can be found at
https://doi.org/10.1007/978-981-16-8531-6_2

Author Index

Printed in the United States
by Baker & Taylor Publisher Services